A PRACTICAL GUIDE
to
STATISTICAL MAPS
and DIAGRAMS

H. C. Truran

Heinemann Educational Books
LONDON

Preface

With the increasing flow of statistics from all quarters it has become more and more necessary to understand and appreciate methods employed to present statistics cartographically and diagrammatically.

It is the aim of this book to outline the main methods available, to describe the construction of each and to discuss the uses, merits or demerits of each. It is concerned, as its title suggests, with the techniques and mechanics necessary in the successful presentation of statistics; it does not deal with the analysis of statistics—a subject which is adequately dealt with elsewhere—nor does it demand from the reader more than an elementary knowledge of mathematics.

Although primarily intended for 'A' level students of Geography, for whom statistical representation usually forms an essential part of the syllabus, it is hoped that this book will be of interest and value to students of economics, history or social studies—to all those, in fact, who may be concerned with the interpretation or production of statistical maps and diagrams.

A comprehensive range of exercises and a set of tables for conversion into and from metric equivalents are included at the end of the book.

Heinemann Educational Books Ltd
LONDON EDINBURGH MELBOURNE AUCKLAND TORONTO
KINGSTON HONG KONG SINGAPORE KUALA LUMPUR
IBADAN NAIROBI JOHANNESBURG
LUSAKA NEW DELHI

ISBN 0 435 34895 7
© H. C. Truran 1975
First published 1975
Reprinted 1977

Published by Heinemann Educational Books Ltd
48 Charles Street, London W1X 8AH
Filmset by Filmtype Services Limited, Scarborough
Printed in Great Britain by
Fletcher and Sons Ltd., Norwich

Contents

Preface   ii
List of Figures   iv
Introduction   1

One: Classification   3
Two: Statistical Graphs   5
  1. Line and curve graphs:
      (i) simple   6
      (ii) group or comparative   7
      (iii) compound   8
      (iv) divergence   8
  2. Bar graphs:
      (i) simple   10
      (ii) group or comparative   11
      (iii) compound or divided   13
      (iv) divergence   13
  3. Age and sex pyramids   14
  4. Dispersion graphs   16
  5. Semi-logarithmic graphs   18
  6. Circular graphs   20
Three: Statistical Charts and Diagrams   22
  1. Divided circles (pie charts):
      (i) simple   22
      (ii) proportional circles and semi-circles 24
  2. Divided rectangles:
      (i) simple   26
      (ii) compound   27
  3. Repeated symbols   29
  4. Proportional (i) circles   30
            (ii) squares   32
            (iii) cubes   34
            (iv) spheres   35
  5. Graduated range of symbols   36
  6. Wind roses:
      (i) simple   38
      (ii) compound   39
Four: Statistical Maps   40
  1. Dot maps   40
  2. Isoline maps   44
  3. Shading (choropleth) maps   45
  4. Flow maps   48
Exercises   50
Appendix : Metric Conversion Tables   59

List of Figures

Fig. 1   Simple Line or Curve Graphs
(a)   Mean monthly temperature of London.   6
(b)   Mean monthly temperature of London, plotted for
      two years.   6
(c)   Growth of population of England and Wales
      1851–1971.   6
Fig. 2   Group (Comparative) Line Graphs
(a)   Principal crops (by area) of U.K. 1963–1967.   7
(b)   Kenya: mineral output by value 1964–68.   7
Fig. 3   Compound Line Graphs
(a)   Principal crops (by area) of U.K. 1963–67.   8
(b)   Kenya: mineral output by value 1964–68.   8
Fig. 4   Divergence Line Graph
Mauritius: production of cane sugar (tonnage) 1961/2–
1968/9; comparison of annual totals with average for
that period.   9
Fig. 5   Simple Bar Graphs
(a)   Mean monthly rainfall, Bombay.   10
(b)   Mean monthly rainfall, Bombay, plotted for two
      years.   10
(c)   Combined temperature (curve) and rainfall (bar)
      for Jerusalem.   11
(d)   World production of tin, 1968.   11
Fig. 6   Group (Comparative) Bar Graphs
(a)   Principal crops (by area) of U.K. 1964–67.   12
(b)   As above, but using overlapping bars.   12
Fig. 7   Compound Bar Graph
Principal crops (by area) of U.K. 1963–67.   13
Fig. 8   Divergence Bar Graph
World production of cocoa (long tons) compared with
world grindings of cocoa 1964–70.   14
Fig. 9   Age and Sex Graphs
(a)   Age structure of the population in selected
      countries: Sri Lanka (Ceylon), Denmark and
      Kenya.   14
(b)   Superimposed pyramids: population of U.S.A.,
      1950 and 1960.   15
(c)   Age distribution of U.K. population by bar graphs.   15
Fig. 10   Dispersion Graphs
(a)   Monthly rainfall figures for Kakira (Uganda)
      1941–70.   17
(b)   Rainfall dispersion graph, Kakira (Uganda),
      1941–70.   17
(c)   Median and quartiles.   17
Fig. 11   Semi-logarithmic Graphs
(a)   Population growth of towns A and B (simple line
      graph).   18
(b)   Population growth of towns A and B (semi-
      logarithmic graph).   18
(c)   Oil production: Africa by countries 1956–68.   19
Fig. 12   Circular Graphs
(a)   Temperature and rainfall: Jerusalem.   20
(b)   Wheat-growing in the Prairies.   20
(c)   Cocoa-growing in Ghana.   20
(d)   Rainfall and crop cultivation: West Teso (Uganda).   21
Fig. 13   Simple Divided Circles (Pie Charts)
(a)   Percentage chart.   22
(b)   Cocoa: world production by countries, 1968.   23
(c)   Ghana: exports (by value), 1967.   23

(d)   Denmark: distribution of population by occupation,
      1960 and 1967.   24
Fig. 14   Proportional Divided Circles
(a)   Semi-circular. World production of commercial
      and passenger vehicles by countries, 1968.   25
(b)   Racial composition of provinces of South Africa,
      1970.   26
Fig. 15   Divided Rectangles
(a)   Australia: area of states, 1971.   27
(b)   Land use in six European countries, 1968.   27
Fig. 16   Repeated symbols
(a)   Zaire: products, 1972.   28
(b)   Denmark: fishing, 1968.   29
Fig. 17   Proportional Circles
(a)   Four methods of constructing proportional circles.   30
(b)   Population of major towns of Netherlands, 1969.   31
Fig. 18   (a), (b) Methods used in drawing key for pro-
          portional circles.   32
Fig. 19   Proportional Squares
(a)   South Africa: tonnage of goods handled at ports,
      1970.   33
(b)   Switzerland: population of major cities and urban
      agglomerations, 1969.   33
Fig. 20   Proportional Cubes
(a)   Comparison of different methods of representation:
      linear, proportional squares, proportional cubes.   34
(b)   West Africa: groundnut production, 1970.   35
Fig. 21   Proportional Spheres
Different methods of drawing spheres.   36
Fig. 22   Graduated Range of Symbols
(a)   Different methods of drawing.   37
(b)   Cornwall: population of towns.   37
Fig. 23   Wind Roses
(a)   Simple monthly wind roses.   38
(b)   Mean annual percentage frequency of wind
      direction and wind speeds.   38
Fig. 24   Dot Maps
(a)   Wales: distribution of population by counties, 1961.   42
(b)   Kenya, Central Province: distribution of
      pyrethrum, tea, pineapples and rice.   43
(c)   Kenya, Central Province: distribution of coffee.   43
Fig. 25   Isoline Maps
(a)   Base-map of thirteen 'counties' showing densities
      at various points.   44
(b)   Base-map as above with isolines added.   44
(c)   Isoline map.   45
Fig. 26   Shading (choropleth) maps
(a)   Different methods illustrated.   46
(b)   Base-map of thirteen 'counties' showing average
      density for each.   47
(c)   Density (shading) map.   47
Fig. 27   Flow map
Frequency of passenger trains per week from Exeter,
April 1970.   49
Fig. 28   Administrative regions of the British Isles.   56
Fig. 29   Southeast England: New Towns.   56
Fig. 30   The Departments of Aquitaine.   57
Fig. 31   Map of Uganda and flight time-table.   58

Introduction

1. One of the most important points in the drawing of any statistical map or diagram is that it must be neat and workmanlike. Cultivate a pride in producing a finished article which, apart from its accuracy and effectiveness, is pleasing to look at. Too often a statistical map*, which has taken a long time to prepare, suffers from an untidiness which spoils the general effect and irritates the reader. A statistical map which is untidy or difficult to read defeats its own purpose.

2. It is essential to examine and, where possible, to emphasise the important points that need to be stressed. By a careful choice of symbol or colour or type of shading†—or, very often, choice of method—it is possible to draw the reader's attention very quickly to the particular point or aspect one wishes to emphasise. It is therefore necessary first to consider what meaning the statistics have, what impression or message one hopes to give to the reader and then to choose the method which will most effectively produce the desired result. Thus the drawing of a statistical map must be regarded not merely as an exercise in representing certain statistics in diagrammatic form but as a means of conveying to the reader in the most effective manner the relative importance of the individual statistics—of bringing to life what would otherwise appear as a lifeless collection of facts and figures.

3. There are often two or more methods of representing the same statistics; usually one method or combination of methods will give a better, clearer, impression than the others. It is necessary, therefore, to analyse the purpose, to compare the available methods and to choose the method which is most informative, incisive and effective.

4. Remember that, as in the case of a topographical map, the more crowded the statistical map is, the more difficult it becomes to read and understand. It is often preferable to draw two statistical maps if there is too much information to be depicted on one.

5. Every statistical map or diagram must have a title and, where necessary, a key and a scale; without these it becomes meaningless.

6. One of the most useful methods of preparing statistical maps is to draw them on tracing paper for use as an overlay. Providing that the base maps are on the same scale comparisons can readily be made in the case of such things as density of population with communications or of, for example, cattle or sheep with relief, distribution of crops with soils and so on. The same principle may be applied also to graphs. By the use of such overlays it is possible to avoid what would otherwise be an overcrowded statistical map on which comparisons and inter-relations would be very difficult to appreciate.

7. Remember that the appreciation of a quantity represented by a linear symbol is easier than when represented by an areal symbol and that the areal symbol in turn presents less difficulty than a volume symbol. A line 2 cm long is obviously twice as long as a line 1 cm long and the ratio between the quantities that each line represents is clearly 2:1. In the case of circles with diameters of these measurements the ratio becomes 4:1 and in the case of spheres 8:1. The assessment of relative values as represented by spheres in particular is extremely difficult: the volume of a tennis ball, for example, is approximately $3\frac{1}{4}$ times that of a golf ball; that of a football approximately 150 times.

8. Graphs drawn for the purposes of comparison must be drawn on the same scale.

9. Basic equipment should include a good ruler, marked in millimetres and inches, a soft rubber, pencil (HB and either H or 2H), an assortment of coloured pencils (with a selection, if possible, of shades of the same colour), a pair of compasses, dividers, protractor (a percentage protractor is a useful addition for drawing pie charts) and a set square (for the drawing of vertical lines). Extra equipment could include an ink compass, parallel ruler, coloured inks and, if desired, stencils for lettering purposes.

10. Do not attempt to draw straight lines without

---

* The term 'statistical map' is used hereafter to include statistical diagrams and graphs.

† A distinction is made throughout between colouring and shading. The term 'colouring' is used here to indicate the use, by pencil or colourwash, of a solid colour or shade or tint of that colour; 'shading' or 'line-shading' indicates the drawing of lines, vertically, horizontally or diagonally (usually in pencil or ink), instead of solid colouring.

the aid of a ruler—for example, when shading density (choropleth) maps with parallel lines. Similarly, in the case of a key, do not draw 'boxes' with free-hand lines.

11. The use of coloured pencils or inks (not wax crayons) is of great value. Avoid the use of pale yellow which is difficult to see in artificial light. Generally speaking, different shades of the same colour are to be preferred to different colours, although striking and significant aspects can sometimes be emphasised by using a different colour; even so, the choice of colours must be such as to avoid a garish, patchwork appearance. The colouring of, for example, pie charts or compound bar graphs, using sharply contrasting colours in juxtaposition can produce a visually interesting and striking effect. But in the case of statistical maps such as those showing density, where the impression of gradual transition from area to area must be given, colours, colour shades or shading must be carefully chosen to avoid the impression of 'jumping' from one area into another, each separated from the other by some sort of barrier or fence. The attention of the reader can, of course, be intentionally and quickly drawn to an area of, for example, very dense population by the choice of a bright colour or a type of shading that stands out from the rest. It is advisable, too, not to leave any area blank, that is, without colour or shading, as this gives the impression of a non-productive, deserted or desert area. When using colours it is essential to strive for an overall appearance and to avoid streaks and patches of darker or lighter colour. Time spent in practising with coloured pencils is not wasted: when colouring larger areas it will be found more advantageous to work with a circular rather than a side-to-side or to-and-fro action. Use a coloured pencil lightly at first; a 'heavier' colour can be obtained by going over the area a second time.

12. In order to maintain a sharp point on pencils it will be found useful to keep a small piece of fine sand- (glass- or emery-) paper at hand —a light rub on this will quickly produce a sharp point.

13. Information given in writing on the statistical map is best done by either printing or by block capitals. As far as possible all printing should be horizontal for ease of reading; do not compel the reader to turn the statistical map through 360° to understand it or to stand on his head!

14. A check should always be made to make certain that the individual parts of the total do in fact add up to the total given. If the total of the components falls short, provision must be made on the statistical map to represent the missing quantities by 'others' (e.g. Fig. 3(b), page 8).

15. Sometimes the quantities to be represented (for example, on a flow map or a compound bar graph) are so small as to make an assessment of the actual quantities difficult for the reader. In such cases it may help to draw an inset at an enlarged scale. An arrow should point clearly to the inset from that part of the statistical map which is being enlarged and the rate of exaggeration (e.g. ×4) must be clearly indicated and a note added by way of explanation.

16. Cultivate the habit of examining statistical maps wherever they appear—in books, periodicals or newspapers. Make an appraisal of them—are they attractive, easy to read and interpret? Are they honest (do they purposely give a wrong impression)? Are there any techniques employed which you have not seen before but could profitably use in your own work? Is there any other method which would have been more effective? Does it, in short, convey the facts more clearly and more easily than the figures would have done?

17. It is advisable, before embarking on the final product, to make a rough preliminary drawing. If drawn on approximately the same scale (accuracy is not essential at this stage), it will give you a good idea of what the final product will look like and may also bring to light difficulties and problems that will require attention.

18. Exercise your own ingenuity in the construction of statistical maps and diagrams: a variation of a known method, a combination of two or more methods, an unusual application of an 'orthodox' method—any of these, if used correctly, can produce striking or graphic results.

19. It cannot be stressed too heavily that continued practice is essential in acquiring the practical skills and techniques of statistical representation; reading about them is not sufficient.

ONE
Classification

A basic distinction exists between, on the one hand, those statistics which are primarily concerned with the relationship between quantities but do not stress the idea of location or spatial distribution—such as figures relating to temperature and rainfall or variation in production of minerals or crops—and, on the other hand, those statistics which inherently imply some idea of location or distribution—such as details of population density and distribution or movement of goods and traffic.

The first of these two types can be subdivided by taking into consideration the methods of representation available, thus making it possible to distinguish three groups in all (see below). It is important, however, to remember at the outset that these groups are not self-contained or watertight and that very often one method of representing a given set of statistics may be used in conjunction with another method. A statistical graph, for example, of the monthly rainfall at Bombay, drawn primarily to show the relationship between the monthly quantities of rain, may acquire the idea of location if drawn in its correct place on a map of India.

Of the many methods employed in the representation of statistics by diagrams or maps, three main groups, therefore, can be distinguished. They are:

1. STATISTICAL GRAPHS

This group, like Group 2, is basically concerned with the relationship between quantities and does not stress the idea of location. It differs from Group 2, however, in that it comprises those methods of statistical representation which have, as their basis for drawing, a background of squared graph paper; the background may or may not be evident from the actual finished drawing (and, indeed, it is more customary to omit the squared background on the grounds that it detracts from the visual impression) but the horizontal and vertical axes must normally appear as a basic and integral part of the drawing.

2. STATISTICAL CHARTS AND DIAGRAMS

This group comprises those methods which do not depend on squared paper or a map as their basis. As in the case of statistical graphs, they may be used in conjunction with a map for the purpose of defining or emphasising location but they can also be drawn independently, this being their main function.

3. STATISTICAL MAPS

This group includes those methods of statistical representation which, because they stress the idea of location or spatial distribution, make use of a map as the basis of their construction. By their nature they are essentially statistical maps, but they may be used in conjunction with other methods of representation.

The following classification is used in this book:

GROUP 1 Statistical Graphs	GROUP 2 Statistical Charts and Diagrams	GROUP 3 Statistical Maps

1. Line and Curve Graphs
  (i) Simple
  (ii) Group (Comparative)
  (iii) Compound
  (iv) Divergence
2. Bar Graphs
  (i) Simple
  (ii) Group (Comparative)
  (iii) Compound
  (iv) Divergence
3. Age and Sex Graphs
  (Pyramids)
4. Dispersion Graphs
5. Semi-logarithmic Graphs
6. Circular Graphs
	1. Divided Circles (Pie Charts)
	  (i) Simple
	  (ii) Proportional
	2. Divided Rectangles
	  (i) Simple
	  (ii) Compound
	3. Repeated Symbols
	4. (i) Proportional Circles
	  (ii) ,, Squares
	  (iii) ,, Cubes
	  (iv) ,, Spheres
	5. Graduated Range of Symbols
	6. Wind Roses
	  (i) Simple
	  (ii) Compound
		1. Dot maps
		2. Isolines
		3. Shading Maps (choropleth)
		4. Flow maps

TWO
Statistical Graphs

1. LINE AND CURVE GRAPHS

This method of representing statistics is one of the most used and one of the most popular. Its appeal stems from its simplicity both in construction and interpretation and from its versatility. It has a wide variety of uses, ranging from temperature graphs (it should not be used to show rainfall totals), population trends, variations in crop or mineral production, fluctuations in trade and so on.

Although the same in construction, two methods of drawing can be distinguished: one in which the points plotted are joined by a series of straight lines—a line graph (see Figs. 2(a), 2(b), 3(a), 3(b)), the other in which they are joined by a smooth, rounded curve—a curve graph (see Figs. 1(a), 1(b), 1(c)). The basic difference can be illustrated by a comparison of a clinical temperature graph and a mean monthly temperature graph. In the case of the former, the body temperature of a patient is recorded every four hours; the points are joined by straight lines which focus attention on and emphasise the rise and fall in temperature. Mean monthly temperatures, on the other hand, are usually recorded by a curved line, indicating the continuity of the rise and fall in temperature without intentionally drawing attention to maximum and minimum values. Similarly, population trends (see Fig. 1(c)) and readings of pressure are shown by a smooth, curved line to emphasise the basic continuity of change, while statistics relating to crops or minerals (on a monthly or annual basis and often characterised by discontinuity or irregularities in production) are usually represented by a line (see Figs. 2(a), 2(b), 3(a), 3(b)) rather than a curve graph. Apart from its more pleasing appearance, however, and the fact that it avoids giving the impression of sharp, angular movement, the curve graph offers little advantage over the line graph as far as accuracy is concerned: in both types accurate interpolation is rarely possible.

The four main varieties of line and curve graphs are:
  (i) Simple
  (ii) Group (or Comparative)
  (iii) Compound
  (iv) Divergence

1(i) Simple Line and Curve Graphs (Figs. 1(a), 1(b), 1(c))

The following points should be borne in mind:

CONSTRUCTION

a. The horizontal axis is normally used to represent the independent variable, i.e. time (whether in hours, days, months, years or any other period of time), the vertical axis to represent the dependent variable, i.e. quantities or values, sometimes as percentages.

b. The base of the vertical scale should be at zero, the top should be slightly higher than the maximum value to be recorded on it.

c. If drawn on plain paper, it is preferable to draw two vertical axes, one at each end of the horizontal axis. By placing a ruler across both axes reading of values will be simplified.

d. Two different sets of values may be marked on the two axes, provided that the relationship between them is constant, for example, °C and °F, hectares and acres, tonnage and value.

e. There are three positions possible when placing the dots: in the middle of the space between the vertical lines, on the vertical line to the left, on the vertical line to the right. As a general guide it is more satisfactory to place the dot in the middle of the space; this will be found more convenient when constructing a combined temperature (line) and rainfall (bar) graph (Fig. 5(c)).

f. Quantities or values marked on the vertical scale must not 'stand on' the horizontal lines, but written so that they would be bisected by the horizontal line if it were produced.

g. Do not indicate large numbers with long strings of noughts, for example, 100,000, 200,000 and so on, but write, either at the top (preferably) or along the side, the value of the units expressed in figures or words, for example, £ '000 or '0,000 tonnes (or £ thousand, ten thousand tonnes).

h. When plotting, do not use crosses or dots surrounded by a circle to mark values. Only a dot should be used—this should be visible, but not too prominent, when all dots are connected.

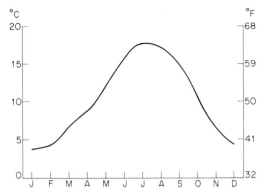

	J	F	M	A	M	J	J	A	S	O	N	D
°C	3·9	4·4	6·7	8·9	12·2	15·6	17·8	17·2	15·0	10·6	6·7	4·4
°F	39·0	39·9	44·0	48·0	53·9	60·0	64·0	62·9	59·0	51·1	44·0	39·9

1(a) Simple line or curve graph: mean monthly temperature of London

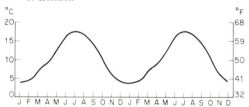

1(b) Simple line or curve graph: mean monthly temperature of London, plotted for two years

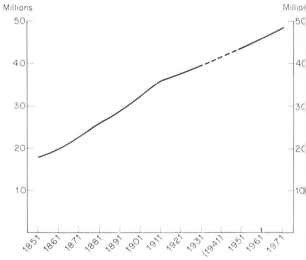

Year of census	1851	1861	1871	1881	1891	1901
Population	17,927,609	20,066,224	22,712,266	25,974,439	29,002,525	32,527,843

1911	1921	1931	1941	1951	1961	1971
36,070,492	37,886,699	39,952,377	—	43,757,888	46,104,548	48,604,000

1(c) Simple line or curve graph: growth of population of England and Wales, 1851–1971

GENERAL

a. Very great care must be taken in choosing both the vertical and horizontal scales. A graph, for example, of mean monthly temperatures for an equatorial station will give a false impression if the vertical scale is exaggerated; a graph representing monsoon rainfall will have less meaning if the vertical scale is unduly reduced.

b. Graphs used for the purpose of comparison must be drawn on the same scale.

c. An explanatory note, written below the graph (for example, '5 mm on the horizontal scale represents one month' or '1 cm on the vertical scale represents 100 tonnes), is neither necessary nor desirable—this information should be obvious from the graph.

d. Squared graph paper forms the basis of drawing, but as the network of squares detracts from the simplicity of a straight or curved line, it is more customary to dispense with the graph paper and to draw the graph on plain paper. To do this, a set square is essential to mark out right angles from the horizontal. Alternatively, the graph may be first drawn on squared graph paper and then traced, marking only the essential co-ordinate lines.

e. The use of a ruler, placed horizontally across the vertical axes, will facilitate reading of the graph.

1(ii) Group (or Comparative) Line Graphs
(Figs. 2(a), 2(b))

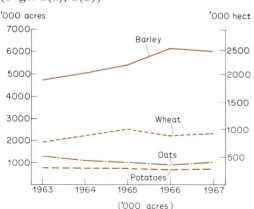

	1963	1964	1965	1966	1967
Wheat	1928	2206	2535	2238	2305
Barley	4713	5032	5395	6130	6027
Oats	1295	1125	1014	907	1012
Potatoes	768	778	741	669	708

2(a) Group line graph: principal crops of U.K.('000 acres), 1963–67. (Source: *Statesman's Yearbook 1968–9*)

As its name implies, more than one line or curve graph can be drawn on the same statistical graph; for this reason it is sometimes known as a multiple line graph. This method is particularly useful for purposes of comparison. The method of drawing is the same as for simple line and curve graphs (above).

CONSTRUCTION

a. Lines and curves should be clearly and easily distinguishable from each other; different colours can be used or different methods of drawing a line can be employed, e.g.

——— , – – – – , · · · · · , –·–·–·– , – · · – · · –.

or any variation of these.

b. The maximum number of lines or curves that can satisfactorily be drawn on one statistical graph is probably four or five, but this depends on the 'spacing' available.

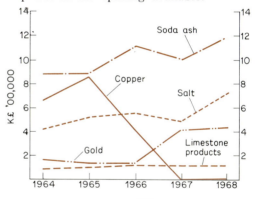

	1964	1965	1966	1967	1968
Soda ash	887,883	895,806	1,183,677	1,093,521	1,203,552
Salt	420,299	528,009	555,121	494,598	731,875
Copper	654,662	868,281	426,270	5 500	14,654
Gold	168,552	142,938	149,490	420,118	447,756
Limestone products	94,380	109,905	126,992	128,420	125,561
Others	185,918	153,180	178,189	175,935	210,467
Total	2,411,694	2,698,119	2,619,739	2,318,092	2,733,865

2(b) Group line graph: Kenya mineral output by value, 1964–68. (Source: *Kenya Economic Survey, 1969*)

GENERAL

a. As the purpose of the graph is to draw attention to comparisons and in order to reduce dependence upon the key, it is sometimes advantageous to write essential information on each line or curve, but this should not exceed one or at the most two words per line.

b. Crossing of lines or curves should be avoided as much as possible, as this may lead to confusion and will increase the difficulty in interpretation. It may be advisable, as in the case of (b) above, to draw two sets of graphs, in which case comparison can be more readily made if one of the lines or curves is repeated on the other statistical graph.

1(iii) Compound Line Graphs
(Figs. 3(a), 3(b))

This method is sometimes used, in the same way as a compound bar graph (see p. 13), to show the individual or component parts of a total, for example, value of minerals or hectarages under certain crops at different dates.

CONSTRUCTION

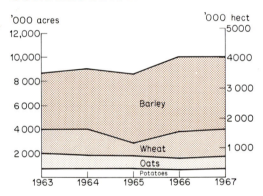

3(a) Compound line graph: principal crops by area of U.K., 1963–67.

a. One graph is first drawn, representing any one component, usually the largest or the one that shows the least fluctuation. A second graph is drawn above this for the next component, adding, for any particular year, its value or quantity to the value or quantity of the 'underlying' commodity for that year.
b. The total value or quantity corresponds with the uppermost graph.
c. Lines cannot cross each other.
d. Colouring or shading of the component parts is of considerable help in interpretation.

GENERAL

a. Because the base of any individual line graph (except the bottom one) is not zero, it must be remembered that only readings between the individual line graphs can be considered when

trying to determine the value or quantity of single commodities.
b. Do not confuse the compound line graph with the comparative (or group) line graph. In the case of the former, all lines represent cumulative totals and, for example, to obtain from the graph in Fig. 3(b) the value of copper output in Kenya in 1966, it is necessary to subtract the total value of soda ash plus salt from the total value of soda ash plus salt plus copper. In the comparative or group line graph (Fig. 2b), however, the value of each individual commodity is calculated from the zero line: totals, therefore, are not cumulative and the value of copper output for any year can be read directly from the graph.

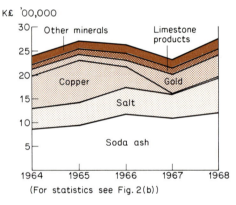

3(b) Compound line graph: Kenya mineral output by value, 1964–68

c. In the case of a compound line graph, the area between the individual graph lines is shaded or coloured; with comparative line graphs the line itself is coloured or otherwise distinguished.
d. Its great disadvantage, therefore, is that, although it represents total values or quantities successfully, it is very difficult to find the value of any one commodity for any particular year or to follow with any accuracy the quantitative rise and fall of any one commodity.
e. A background of closely drawn squares would detract from its appearance.

1(iv) Divergence Line Graphs (Fig. 4)

It is sometimes necessary to show 'plus' and 'minus' values, such as export or import figures over a number of years which are above or below average for that period, profit or loss, increase or decrease in population as a result of migration, fluctuation in crop or mineral production in relation to normal or average conditions. One way of doing this is by a line graph on which values are plotted in relation to a zero line, which can represent either the average of the period chosen or a particular year selected for purposes of comparison. Fig. 4, for example, could be drawn to show the rise or fall in production as compared with any year, e.g. 1967.

CONSTRUCTION

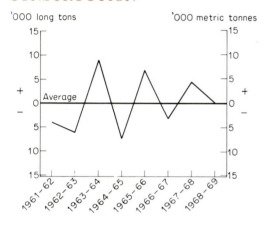

	1961/2	1962/3	1963/4	1964/5	1965/6	1966/7	1967/8	1968/9
'000 long tons	545	524	675	511	654	553	628	587

4 Divergence line graph: Mauritius—production of cane sugar (tonnage) and comparison of annual totals with average for period 1961/62—1968/69. Average annual production for period 585,000 tons.

a. For purposes of interpretation or emphasis the zero line is usually thickened.

b. Great care must be taken in choosing the vertical scale; exaggeration, for example, may suggest a calamitous fall in production of exports or a fictitious 'boom' year.

c. The vertical scale can be in terms of values, quantities or percentages.

d. The horizontal axis representing the independent variable should not be confused with the zero line but should occupy its normal place at the base of the graph.

e. Values above average are calculated by subtracting average from actual production totals and plotted accordingly; 'short-fall' figures are plotted below the zero line. These values may also be plotted as percentage rise and fall.

f. The significance of the zero line should be clearly stated on or below the graph.

GENERAL

a. The graph does not normally show production, exports, population and so on as absolute totals; it shows divergence, positive or negative, from some particular or average condition.

b. Because of the possibility of confusion arising from the crossing of lines, this type of graph is usually drawn for one commodity only; its use is normally restricted to line rather than curve graphs. It should not be used as a compound or comparative line graph.

2. BAR GRAPHS

Like line and curve graphs, the popularity of bar graphs (sometimes called columnar graphs) arises not only from their simplicity of construction but also from their visual appeal. The information they convey is clearly and easily understood, the length of the bars being in all cases proportional to the numbers they represent. In many respects they are similar to line and curve graphs, their construction being essentially the same. But, whereas the line or curve graph emphasises the rise and fall of values, the bar graph, by its representation as and impression of definite quantities, draws attention both to the individual amounts and their relative variations. The effect of a solid bar, coloured or shaded, in contact with its base line is much more concrete and indicative of a definite quantity than a line suspended above its base line and not in contact with it. The line or curve graph gives a qualitative assessment by emphasising the overall variation or rise and fall in values over a period of time; the bar graph, on the other hand, is of a more quantitative nature, drawing attention to the individual amounts that make up that variation. It would be difficult, for instance, to appreciate fully the continuity of the rise and fall of mean monthly temperatures if they were represented by a bar graph.

The four main varieties of bar graphs are:
(i) Simple
(ii) Group (or Comparative)
(iii) Compound (or Divided)
(iv) Divergence

2(i) Simple Bar Graphs (Figs. 5(a), 5(b), 5(c), 5(d))

CONSTRUCTION

a. The horizontal scale usually represents the independent variable, more especially when the time element is of significance, such as when graphing mean monthly rainfall (see Figs. 5(a), 5(b)) or annual production of a commodity. This arrangement, however, is not inflexible (see (d) below).

b. A vertical bar may occupy the space (either completely or partially) between two vertical lines: the vertical line does not bisect the bar. Periods of time, therefore, in this case, are assumed to occupy the spaces and are written, not under the vertical lines, but under the spaces, between the vertical lines.

c. All bars must start at zero; bar graphs drawn for the purposes of comparison must be drawn on the same scale.

d. Bars may be drawn horizontally when the time element is of minor significance, for example, production of one commodity by different countries for a particular year (see

Fig. 5(d)). In such cases, bars are usually arranged in order of magnitude, the largest at the top.

e. When vertical bars are drawn, the time sequence is from left to right.

f. If statistics for any one year are missing, a space should be left to indicate this.

g. The width of the bar is a matter of choice; avoid bars that are too thick or too thin. The value of each bar can be assessed more easily if a space or gap is left between each bar. This also applies to mean monthly rainfall bars which, if separated, give a clearer impression of individual monthly values.

h. As with line and curve graphs, a limited number of co-ordinate lines will make assessment of values easier.

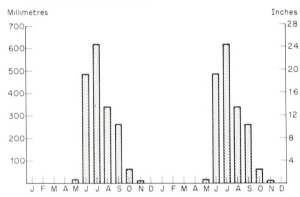

5(b) Simple bar graph: mean monthly rainfall at Bombay plotted for two years

GENERAL

a. The length of the bar, not the width, is the important factor.

b. It is easier to add information on the graph if bars are drawn horizontally, but see (d) above.

c. Bar graphs are often used in conjunction with line or curve graphs (see Fig. 5(c)). This is often done in atlases to present a more complete picture of rainfall and temperature. In such cases the two vertical axes are scaled; one in degrees centigrade and the other in millimetres or centimetres.

d. The bar graph is in many cases interchangeable with a line or curve graph; much depends on where the emphasis is to lie.

e. Mean monthly rainfall figures can also be plotted as a percentage of the mean annual rainfall. This method facilitates calculation of the seasonal distribution.

f. As with line and curve graphs, the bar graph

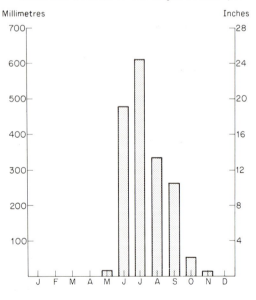

	J	F	M	A	M	J
mm	2·5	2·5	2·5	–	17·8	485·1

	J	A	S	O	N	D	Total
	617·2	337·8	264·2	63·5	12·7	2·5	1808·3

5(a) Simple bar graph: mean monthly rainfall at Bombay

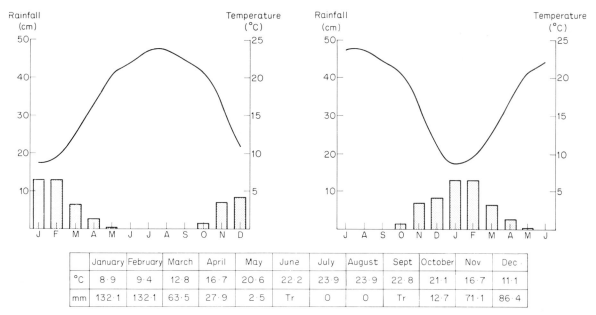

	January	February	March	April	May	June	July	August	Sept	October	Nov	Dec
°C	8·9	9·4	12·8	16·7	20·6	22·2	23·9	23·9	22·8	21·1	16·7	11·1
mm	132·1	132·1	63·5	27·9	2·5	Tr	0	0	Tr	12·7	71·1	86·4

5(c) Simple bar graph in conjunction with curve graph: mean monthly temperature and rainfall for Jerusalem. These two graphs represent the same information

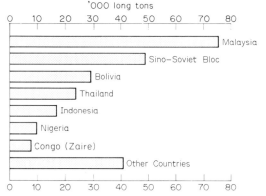

5(d) Simple bar graph: world production of tin ('000 long tons of tin-in-concentrates), 1968. (Source: *Geographical Digest, 1970*)

may acquire the idea of location when super-imposed on a map in its correct place, for example, when showing rainfall or exports.

g. A better idea of the yearly rhythm of rainfall and its cyclic nature will be gained if the horizontal axis is extended to cover two years and statistics for a two-year period are plotted (Fig. 5(b)). Many false impressions concerning both annual distribution of rain-fall (temperature, too) and rainfall regimes in different hemispheres would be avoided if mean monthly rainfall figures were plotted on either an eighteen-month or two-year basis (cf. Fig. 1(b)). The inhabitant of the tropical

savanna or of the monsoon regions is no less interested in the arrival and length of the wet season than he is in the length of the dry season and the shortage of food.

2(ii) Group (or Comparative) Bar Graphs
(Figs. 6(a), 6(b))

As with line and curve graphs, bars may be grouped together for the purposes of comparison. Total production of minerals, for example, for any particular year may be represented by a group of bars standing side by side, each bar referring to one mineral, the total length of the individual bars when added together indicating the total mineral production.

Compared with the group line graph, the grouping of bars gives a better impression of totality and of the individual contribution made by each of the component parts but at the same time does not give an accurate impression of the total; this disadvantage can be overcome to some extent by drawing bars, not in terms of absolute quantities, but as percentages of the total—this involves modification of the vertical scale only.

As is the case with the simple bar graph attention is directed towards quantities rather than to the rise and fall or fluctuations in the commodities being represented. Because of this it is difficult to follow the general movement or trend of any particular component or, indeed, of the total from year to year. Its use, therefore, is dictated in large measure by the direction or degree of emphasis that is intended; if the purpose of comparing trends or movements in components or totals is essential then a group line graph is preferable, but if attention is to be drawn to the actual quantities of the components, a group bar graph will serve the purpose.

This method of representation, therefore, gives a quick visual impression of the relative importance of individual quantities as each bar stands on the zero line, but is of less value when a comparison of totals forms an important part of the interpretation.

CONSTRUCTION

a. The method of drawing group or comparative bar graphs is the same as for a simple bar graph (see p. 10).

b. To give an impression of totality, bars are usually drawn touching each other, i.e. without a gap between them (see Fig. 6(a)), but attention may be drawn to individual components by leaving a small space between the bars. Groups of bars must, of course, be separated from each other.

c. As the time element is usually of considerable importance (see p. 10), group or comparative bars are normally drawn vertically.

d. It is customary to draw the longest bar of the group on the left, proceeding in descending order to the right. The bar representing 'others' is usually placed on the extreme right of the group.

e. The same order of components (usually from left to right) must be adopted for each group of bars.

f. All bars must be of the same width and drawn at right angles to the axis!

g. As in examples elsewhere, reading the graph will be easier if scale numbers are included on both vertical axes.

h. The use of distinctive colour or shading for different components is advantageous. When comparing production over, for example, a period of years, it is essential that the same colour or shading be used to indicate one component throughout.

i. No writing need appear on the bars but a key must be added.

GENERAL

a. There is, of course, no fixed limit to the maximum number of bars that can be grouped together. For ease of reading and for interpretation purposes, however, groups should be composed of not more than 3 or 4 bars.

b. For the purposes of comparisons bars may also be drawn in pairs. Thus a bar representing total production of a certain crop may be drawn beside a bar representing total rainfall that year, price paid to the grower, or yield per hectare; tonnage of exports from a port may be paired with imports or the number of vessels calling at that port, and so on. In such cases the two vertical axes must be scaled in accordance with the two sets of values and a clear indication given both on the axes and in the title.

c. Bars may be drawn 'inside' each other (see Fig. 6(b)), the overlap of the larger component being complete on three sides. But because this tends to imply comparison on an areal basis rather than on a linear basis only (suggesting that the width of the bar has some significance), it is not recommended.

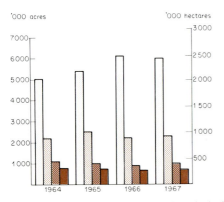

6(a) Group or comparative bar graph: principal crops (by area) of U.K., 1964–67

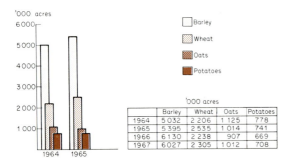

6(b) Group of comparative bar graph: as Fig. 6(a) but using overlapping bars

'000 acres	Barley	Wheat	Oats	Potatoes
1964	5 032	2 206	1 125	778
1965	5 395	2 535	1 014	741
1966	6 130	2 238	907	669
1967	6 027	2 305	1 012	708

2(iii) Compound (or Divided) Bar Graphs (Fig. 7)

A compound bar graph, similar to the compound line and curve graph, is drawn by sub-dividing one bar into its component parts. The total length of the bar represents the total value of the component parts which are shown as sub-divisions (see Fig. 7).

When compared with the compound line graph, the compound bar graph shows totals more clearly but conveys an impression of discontinuous production, while the value of individual components is more difficult to assess. The compound line graph, however, draws less attention to totals, presents the same difficulty in calculating components but gives the impression of continuity, of rise and fall in total production or quantities.

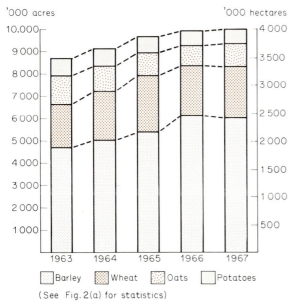

7   Compound (or divided) bar graph: principal crops of U.K. ('000 acres), 1963–67

CONSTRUCTION

a. As the compound is to the simple line or curve graph, so the compound bar is to the simple bar graph and the same general rules are followed.

b. The vertical axes may be scaled in values or in percentages; if two or more compound bar graphs are drawn for purposes of comparison on a percentage basis, they will be of the same linear extent whether vertical or horizontal.

c. In deciding whether to draw the compound bar vertically or horizontally the same principle applies (see p. 10): if representation or comparison over a period of time is the important element, bars should be drawn vertically; when dealing with, for example, production or items of export from several countries for one particular year, bars may be drawn horizontally (see Fig. 5(d)).

d. When drawn vertically the bars may be sub-divided in descending order of size (the smallest at the top); 'others' is usually represented by the sub-division farthest from the zero line. Alternatively the component that shows the least variation may be placed nearest the zero line; sometimes its position may be taken by the component to which it is desired to draw special attention.

e. As the component nearest to the zero line is the only sub-division which can easily be read with any accuracy, it is of advantage to continue the division lines across and between the vertical bars. This also makes it easier to follow the rise and fall of individual values (see Fig. 7), but it must not be assumed that this line indicates a regular movement of rise or fall.

f. Colouring or shading of the components improves the appearance and assists interpretation. Over a period of time the same colouring or shading must indicate the same component.

GENERAL

a. Although the length of bars—and hence of totals—is easily and quickly compared (and in this respect it has an advantage over the group or comparative bar graph) it is difficult to assess the value of any one component or to trace its fluctuation over a period of time.

b. In the case of bars drawn on a percentage basis remember that, although the length of the bar does not vary from year to year, total values almost certainly do.

2(iv) Divergence Bar Graphs (Fig. 8)

The construction of a divergence bar graph is so similar to the divergence line graph that little need be said about it here. Its advantages and disadvantages when compared with its counterpart in line graphs have already been discussed; details concerning the drawing of the bars have been described.

CONSTRUCTION

a. The zero line must be clearly indicated, usually by thickening.

b. As the bar, unlike a line, stands on the zero line, it is possible to draw two zero lines separated by a narrow space in which the horizontal scale can be written. Each zero line must be clearly labelled, but as this method interrupts the scale of the dependent variable, there is little advantage to be gained from it. The horizontal scale, in fact, is best written at the bottom (and top) of the graph.

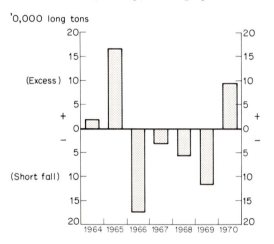

	1963–4	1964–5	1965–6	1966–7	1967–8	1968–9	1969–70
World production '000 long tons	1202	1505	1213	1337	1340	1227	1416
	1964	1965	1966	1967	1968	1969	1970
World grindings '000 long tons	1183	1339	1387	1369	1397	1344	1323

8   Divergence bar graph: world production of cocoa (long tons) compared with world grindings of cocoa, 1964–70. Notice that on this graph annual production figures relate to the year previous to the year of grinding. (Source: *Barclays Bank Annual Review*)

c. The vertical axes must be scaled, both above and below the zero line, the upper part for positive, the lower for negative values.

GENERAL

a. Compound or group bars should not be used to show divergence; divergence bar graphs usually employ simple bars only.

3. AGE AND SEX GRAPHS OR PYRAMIDS (Figs. 9(a). 9(b), 9(c))

With the increasing interest being shown in demographic studies this type of bar graph has become much more widely used.

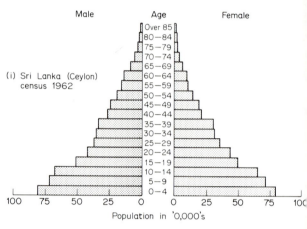

(i) Sri Lanka (Ceylon) census 1962

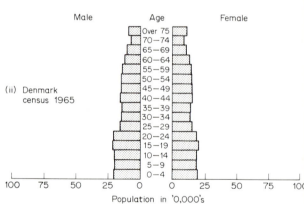

(ii) Denmark census 1965

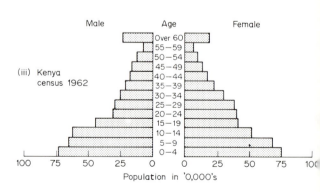

(iii) Kenya census 1962

9(a) Age and sex graph or pyramid: age structure of the population in selected countries. Notice that the upper age groupings are not the same in all three examples. (Source: *U.N. Demographic Year Book*)

Bars drawn horizontally are built up, the length of the bar corresponding to the size of the age group. Ages are usually based on five-year (quinquennial) periods (0 to 4 years, 5 to 9, 10 to 14, and so on), the youngest age group forming the base of the graph. Males are represented on one side of the graph, females on the other. The resulting graph takes the form of a pyramid and is, in effect, a group or comparative bar graph drawn horizontally; the same figures can, in fact, be represented, but less impressively, by a compound bar or a comparative bar group (Fig. 9(c)).

CONSTRUCTION

a. The horizontal axis is scaled either in absolute values or as percentages; notice that the independent variable occupies the vertical axis.

b. In calculating percentages two methods are possible: either, the individual male (or female) groups can be calculated as percentages of the total male (or female) population; or each group may be calculated as percentages of the total population (male and female). In the first case, the shape of the resulting pyramid will be the same whether the absolute or percentage method is used—modification of the horizontal scale only is involved. The second method will, however, produce a different shape, as each bar, because it represents a different percentage, will be of a different length when using this method. In either case, a clear indication must be given, in the title or on the graph, of the method used.

c. Males are usually represented on the left, females on the right. This also must be indicated on the graph.

d. The male and female sections of each bar must be separated; in the space provided it is customary to print the vertical scale in terms of age groups.

e. The horizontal scale must be printed on both sides, i.e. on the male and on the female side.

f. When used for purposes of comparison, the graphs must, of course, be drawn on the same scale.

g. Care must be taken in choosing the scales; an exaggerated vertical or horizontal scale will produce a graph that is either too narrow and elongated or one that is too wide and flat.

GENERAL

a. This method of representation gives a clear picture or summary of population composition that is visually attractive and effective. Its value is enhanced when two or more such graphs are drawn for purposes of comparison; this is particularly true, for example, when comparing the age structure of 'new' towns and older settlements, of industrial towns and health resorts or of developing and developed countries (Fig. 9(a)).

b. Pyramids may also, for purposes of comparison either in terms of time or location, be superimposed. Care must be taken in the construction to ensure that two different methods of shading or colouring are employed. A small space can also be left between each bar, the 'outline' of the superimposed graph being continued across these spaces (Fig. 9(b)).

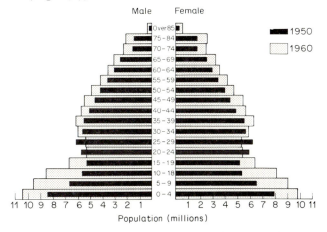

9(b) Superimposed pyramid: comparative age and sex structure of U.S. population, 1950 and 1960. Notice the population loss in the 20–29 age groups since 1950.

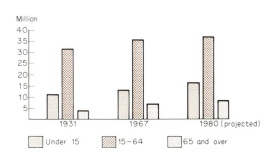

9(c) Bar graphs: age distribution of U.K. population for selected years. (Source: *Central Office of Information*)

c. Pyramids may be drawn to represent growth of total population over a period of years. Each bar then represents, not age groups divided into male and female sections, but the absolute decennial totals as given, for example, in census enumerations. In such cases the vertical axis represents ten-year periods, the horizontal scale indicates absolute totals. Bars may be arranged as a pyramid (usually with the earliest date at the top) and may also be sub-divided to show, for example, rural and urban population, agricultural and industrial workers and so on.

4. DISPERSION GRAPHS (Figs. 10(a), 10(b), 10(c))

Although there are many types and many refinements of dispersion graphs, only two uses need concern us here.

One of its uses is concerned with plotting the dispersion, frequency or spread of some particular phenomenon over a period of time, for example, amount of rainfall by months over a period of years (see Figs. 10(a), 10(b), 10(c)), or by hours over a period of days or months. It need not, of course, be confined to the recording of rainfall; when the principle is understood no doubt many other uses will occur to you—for example, the number of visitors to a museum at specific hours of the day over a period of a month or a year, the number of vehicles passing a particular point at certain times of the day and so on. The graph draws attention to areas of concentrated activity, absence of activity and the overall pattern or spread and is a useful method of representing statistics in that it is easy to construct and visually informative.

A dispersion graph may also be used in determining critical values when it is necessary to subdivide a range of numbers into significant groupings. By plotting the whole range on a dispersion graph such groupings will often be revealed. This method can be of particular value in deciding a graduated range of symbols (see page 36).

CONSTRUCTION

a. One dot represents one value which is plotted against the vertical scale. The dot itself indicates an occurrence or a fixed number of occurrences and the value or quantity it represents can be interpreted by reference to

the vertical scale. Dots, therefore, are usually of uniform size, although it is possible to introduce different colours or shapes to denote occurrences of a specific character.

b. The independent variable is usually represented on the horizontal axis.

c. The vertical axis is normally scaled from zero.

GENERAL

a. As compared with a bar graph showing the mean monthly rainfall over a period of years, a dispersion graph, using the same statistics, indicates all occurrences, including extremes over the same period. From it one can assess the variability in fall for each month and, hence, to a great extent, its reliability. A long period of records (preferably thirty to forty years) is therefore necessary, as impressions gained from a short-term period can be misleading.

b. The graph does not instantly convey average conditions, but illustrates the pattern of occurrences over a period of time, indicating groupings or 'bunchings' of such occurrences.

c. As with days of calm on a wind-rose (see page 38), it is helpful when plotting monthly rainfall figures to indicate (below the horizontal scale under the appropriate column) the number of occasions in each month when no rain fell—this obviates the necessity to count the number of dots.

d. Using statistics provided in Fig. 10(a), mean monthly rainfall figures can be calculated for any particular month by adding the total values for that month over a period of thirty years and dividing by the number of values (i.e. thirty). The mean monthly rainfall for January, for example, is thus

$$\frac{140 \cdot 70 \text{ cm}}{30} \text{ or } 4 \cdot 69 \text{ cm};$$

for November the figure is

$$\frac{419 \cdot 93 \text{ cm}}{30} \text{ or } 13 \cdot 99 \text{ cm}.$$

But it can be seen that, although the mean value thus obtained takes into account all the figures, it can be unduly influenced by extreme or 'freak' values, as in November when, in two years, rainfall in that month exceeded 35 cm and in three years was less than 2·5 cm. In such cases, greater significance may be attached to the *median* rather than

	JAN	FEB	MAR	APR	MAY	JUNE	JULY	AUG	SEPT	OCT	NOV	DEC	Total
	cm												cm
1941	4·42	8·59	8·70	22·99	17·40	11·30	1·90	9·55	4·95	16·82	25·40	10·49	142·51
2	5·72	2·47	20·10	24·20	24·80	2·14	0·31	13·70	6·33	8·20	12·60	8·00	128·57
3	0·97	10·20	2·28	20·60	13·80	3·85	5·55	12·90	9·75	8·35	1·85	2·15	92·25
4	4·95	4·35	13·30	28·20	12·45	3·66	7·31	18·85	7·38	7·80	12·12	7·28	127·70
5	2·18	0·74	4·85	3·65	22·00	12·28	8·64	11·23	6·34	6·90	16·70	1·55	97·06
6	1·78	—	6·62	22·80	16·08	8·52	2·24	17·50	15·60	9·32	11·50	5·65	117·61
7	7·99	5·70	11·58	15·38	24·20	8·23	17·80	8·23	21·90	5·08	1·52	4·48	132·08
8	4·70	5·00	10·81	16·60	12·21	4·80	8·99	10·58	10·38	8·48	11·10	3·67	107·30
9	0·08	2·34	5·86	16·90	8·65	3·15	5·42	14·80	9·45	7·75	13·80	18·70	107·89
1950	0·23	2·95	11·99	17·70	13·80	6·90	11·97	11·68	7·80	5·33	2·31	15·44	108·10
1	9·00	4·09	16·60	18·00	15·99	6·10	4·85	11·10	16·80	13·50	17·01	32·00	165·04
2	1·88	10·41	21·50	21·30	11·60	4·25	3·35	9·68	12·08	5·41	11·70	—	93·16
3	8·40	0·36	7·05	15·99	8·25	12·30	1·37	5·23	9·70	8·85	13·40	4·25	85·15
4	1·96	1·40	3·77	14·60	14·08	6·48	8·00	8·51	9·50	9·32	8·70	4·85	91·17
5	7·50	6·70	12·00	20·70	12·10	8·48	6·95	16·10	9·63	10·08	8·52	13·10	129·86
6	10·20	0·92	13·80	19·21	10·80	10·00	4·67	11·20	6·90	13·81	12·21	5·85	118·57
7	7·05	4·07	6·35	12·80	10·38	7·90	9·75	8·10	2·59	4·26	17·40	6·13	96·77
8	2·68	11·23	8·81	16·40	8·49	13·60	9·31	8·81	5·10	15·00	5·02	11·84	116·28
9	8·20	10·68	4·96	16·80	6·30	6·39	8·98	11·60	12·00	12·80	21·20	3·20	133·11
1960	8·66	7·03	19·19	21·10	7·10	0·18	6·77	2·59	10·89	13·50	5·99	3·45	106·45
1	1·45	2·67	21·60	19·40	12·50	7·22	5·75	13·00	21·40	29·30	47·30	13·70	195·29
2	4·55	1·39	11·28	16·60	14·80	9·19	8·65	5·50	13·79	30·56	9·60	1·89	127·80
3	9·14	5·99	12·30	20·18	12·67	2·61	1·57	4·55	7·90	10·59	20·55	11·60	109·65
4	3·19	14·61	13·58	16·20	11·81	10·90	8·10	5·91	6·92	13·10	11·32	11·22	126·23
5	3·05	2·69	8·35	15·41	3·15	2·27	8·10	9·30	12·82	14·10	14·20	10·00	103·44
6	4·04	13·60	13·35	14·00	5·39	8·88	4·03	8·15	16·00	13·20	7·00	2·00	109·84
7	1·42	3·43	11·65	16·10	15·96	10·85	11·14	5·82	12·96	14·40	36·30	6·35	146·38
8	—	9·67	12·60	17·60	15·90	7·80	6·35	7·58	4·26	12·00	12·25	12·80	128·81
9	7·88	14·74	7·56	13·91	15·61	1·19	3·30	3·28	12·16	9·83	19·32	2·31	121·71
1970	11·43	6·45	23·84	15·79	19·37	6·72	2·70	7·55	5·52	8·89	12·04	2·85	124·15

10(a) Monthly rainfall figures for Kakira, Uganda, 1941–70

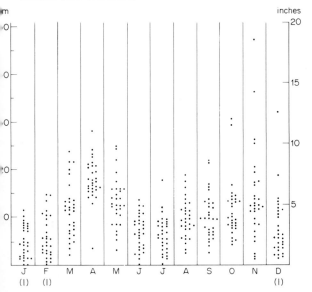

10(b) Dispersion graph: rainfall for Kakira, Uganda, 1941–70. Figures in brackets indicate the number of years when no rain fell in that month.

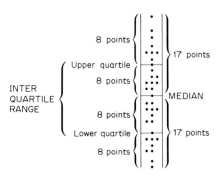

10(c) Dispersion graph median and quartiles. For explanation see page 18.

to the mean. The figures are first arranged in ascending (or descending) order of magnitude (as in the dispersion graph, Fig. 10(b)); the median is that value which has as many dots above it as it has below it. A period of thirty-five years would provide a median which is the eighteenth point (i.e. having seventeen points above it and seventeen below); for a thirty-year period the median would be the value lying mid-way between the fifteenth and sixteenth points. To decide which point to take, add one to the number of points and then divide the result by two. Similarly, the *lower quartile* has one-quarter of the points below it and the *upper quartile* one-quarter of the points above it (see Fig. 10(c)). This time, add one to the total number of points and divide the result by four. Thus, for a thirty-five year period, the lower quartile would be the ninth point from the bottom and the upper quartile the ninth point from the top, while for a thirty year period the eighth point down and the eighth point up can be taken as the nearest values. The median thus gives equal weight to every occurrence, regarding all occurrences as being of equal importance. The values between the upper and lower quartiles—the *inter-quartile range*—therefore do not give undue weight to 'freak' values and may be regarded as a good indication of central 'tendency' or expectation.

5. SEMI-LOGARITHMIC GRAPHS
(Figs. 11(a), 11(b), 11(c))

It is sometimes necessary to represent not only absolute quantities but also rate of change—that is, rate of increase or decrease within any given set of statistics. On a simple line graph, for example, the increase in population of a small town (A) which grew from, say, 5,000 to 10,000 over a period of 50 years, would be represented by a line very different in slope from a larger town (B) of 25,000 which increased over the same period to 50,000 (see Fig. 11(a)). Although the increase in absolute terms of the larger town was 25,000 compared with 5,000 on the part of the other, the rate of increase in both was, in fact, the same: both towns doubled their population within the same period. Similarly, the rate of increase in total exports or commodities (see Fig. 11(c)), production per man-hour as a result of mechanisation, factory output and so on, is

often of more importance than the actual amount of increase. Representation by a line graph can often be misleading in such cases.

On a semi-logarithmic graph the vertical axis is not based on an arithmetical scale (although the horizontal scale is) but on a logarithmic scale; that is, the spacing between the numbers is not proportional to the numbers themselves

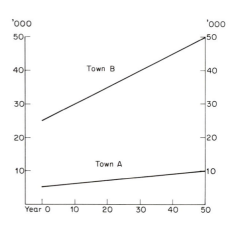

11(a) Simple line graph: growth of population of two towns

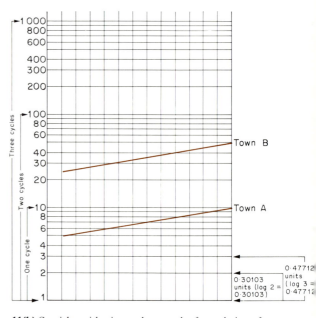

11(b) Semi-logarithmic graph: growth of population of two towns drawn on three-cycle semi-logarithmic graph paper with details of its construction. For purposes of comparison with Fig. 11(a) the growth in population (rate of increase) of the two towns is plotted on the same horizontal scale.

(which would result in even spacing) but to the logarithms of these numbers. Thus spacing of numbers becomes closer as one proceeds along the vertical axis (see Fig. 11(b)). Log 1 is 0·00000, Log 2 is 0·30103, Log 3 is 0·47712 and Log 10 is 1.00000. Therefore if the base of the vertical scale, or the beginning of a 'cycle', be taken as Log 1 (0·00000), then Log 2 (= 0·30103) will be represented by a point on the vertical scale which is 0·30103 of the vertical distance to Log 10 (= 1·00000), Log 3 by a point 0·47712 of the distance to Log 10 and so on. It is possible to draw one's own vertical scale on this basis, using logarithmic tables, but it is a difficult and tedious task and it is customary to buy logarithmic graph paper specially ruled. Single cycle paper can be scaled from 1 to 10 or from 10 to 100 or from 0·1 to 1·0, according to requirements. Any number of cycles may be employed but the top

and bottom of each cycle must be a multiple (or decimal) of 10. Zero, therefore, can never be reached on a semi-logarithmic graph. Thus a 4-cycle graph paper can be scaled from 0·1 to 1,000 or from 100 to 1,000,000. Each complete cycle is represented by the same vertical distance and it will be noticed that there is the same vertical distance between any number and double that number (for example, 1 and 2) in one cycle, as there is in any other cycle (for example, 100 and 200, 0·1 and 0·2 etc.). Thus— and this is the characteristic from which it derives its difference and its value—similar increases (double, treble, and so on) in any part of the graph or graphs are represented by similar slopes; this is not true in the case of line graphs.

CONSTRUCTION

a. As it is customary to buy semi-logarithmic graph paper already ruled, it is not necessary to give details concerning construction here, although it can be attempted using the information given above. The vertical axis is described in terms of the number of cycles (for example, one-cycle, three-cycle paper) while the description of the horizontal scale (for example, 5 mm) refers to the spacing between the vertical lines.

GENERAL

a. If used solely for the purpose of graphing a large range of numbers (and it should be stressed that this is not its real purpose), it must be remembered that the graph then shows not absolute increases in the same way as a simple line graph (although absolute totals can be calculated from the vertical scale), but *rate of change*, either increase or decrease. It should, therefore, not be used on a direct comparative basis with a line graph.

b. Because its construction is different from other types of graph, it is essential that the vertical scale be clearly marked and that the title include the description 'semi-logarithmic graph'.

c. The difference between a semi-logarithmic and a logarithmic graph lies in the fact that in the latter both the vertical and the horizontal scales are logarithmic, but these need not concern us here.

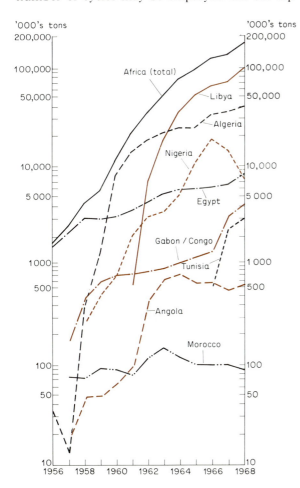

11(c) Semi-logarithmic graph: oil production of Africa by countries, 1956–68

6. CIRCULAR GRAPHS (Figs. 12(a), 12(b), 12(c), 12(d))

Because of their resemblance to the face of a clock or to the lines of longitude radiating from the Pole, circular graphs are sometimes known as clock graphs or polar graphs. They should not, however, be confused with the divided circle or pie chart (see p. 22). The analogy between the twelve months of the year and the twelve hours of the clock face adds attraction to the use of this type of graph when the idea of continuity or progress throughout the year and from year to year needs to be stressed. Although on first consideration it appears to have many advantages, it suffers from several serious disadvantages which limit its effectiveness and its use (see below). It is included here under the heading of statistical graphs because it is basically a simple line or bar graph, the ends of which have been 'bent' to form a circle, in the same way that a world map drawn on a cylindrical projection can be 'bent' into a circle to show the equator as a continuous line. Thus the horizontal axis of the line graph becomes the circumference of the circular graph, its vertical axis one of the radii (usually '12 o'clock').

CONSTRUCTION

a. The circle can be of any convenient size.
b. Twelve equi-angular radii are drawn, 12 o'clock usually representing January, the other months following in clockwise order (see Fig. 12(a)).
c. As a clock graph is most frequently used to represent climatic statistics, radii are scaled in °C, the scale being indicated on either the 12 o'clock or 6 o'clock radius. Points plotted are then joined as a continuous curve.
d. Bars can also be drawn along the radii to indicate mean monthly rainfall totals. In order to avoid congestion at the centre of the circle zero is normally represented as a circle. When both temperature and rainfall are graphed, the appropriate scales can be shown one on either side of the 12 or 6 o'clock radius.
e. A circular graph can be used to advantage to show seasonal cultivation or activities. One way to do this is to draw two concentric circles; the outer circle can be used to show mean monthly rainfall either by bars or by arcs of circles which occupy the spaces between the

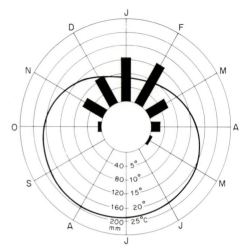

12(a) Circular graph: mean monthly temperature and rainfall for Jerusalem (see Fig. 5(c) for the statistics)

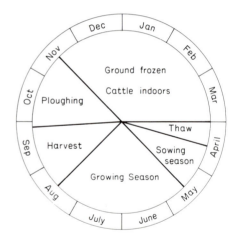

12(b) Circular graph: wheat-growing in the Prairies

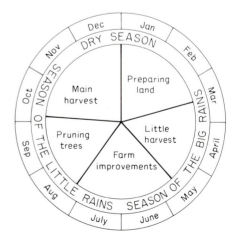

12(c) Circular graph: cocoa-growing in Ghana

radii (see Fig. 12(d)), the 12 or 6 o'clock radius being scaled for amounts as before. The inner circle can be used to show months of planting, growing and harvesting or any seasonal agricultural work. Care must be taken, however, to avoid trying to convey too much information on such a graph.

f. Printing on a circular graph is generally a problem; it is essential to remember that the reader must not be compelled to twist either the graph or his head more than is absolutely necessary! Do not continue writing around the graph, turning it through 360° as you write—this will result in half the writing being upside down. If only initials of months are used, as in Fig. 12(a), there is no difficulty; if other writing is necessary the best solution would appear to be as in Fig. 12(b).

GENERAL

a. This method of representing temperature and rainfall sets out to convey an impression of cyclic development or continuous progression, but it cannot be described as completely successful. The main reason for this is that all statistics relating to temperature and rainfall have to be calculated by reference to a scale which is particularly difficult to follow around the graph; only values along the 12 or 6 o'clock radius can be readily assessed.

b. The fact, too, that the bars representing rainfall values do not stand on a straight line and are not parallel to each other, makes comparison of their values more difficult.

c. Generally speaking, it is less successful in representing monthly values and fluctuations than a combined line and bar graph constructed for an eighteen-month or two-year period.

d. It can be of value, however, as a seasonal cultivation chart where exact values relating to activities are not really possible or necessary.

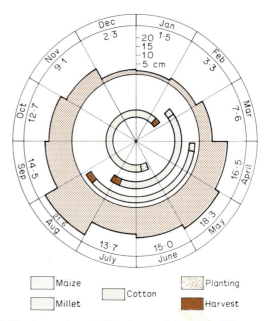

12(d) Circular graph: rainfall and crop cultivation for West Teso (central Uganda). Figures indicate mean monthly rainfall in centimetres. (Adapted from Hickman and Dickens, *Lands and Peoples of East Africa*, Longman 1960)

THREE
Statistical Charts and Diagrams

1. DIVIDED CIRCLES (PIE CHARTS)

Unless one uses circular graph paper for their construction, divided circles are normally drawn on plain paper and thus are more in the nature of charts or diagrams.

Together with line and bar graphs, divided circles (often called pie charts from their obvious resemblance to the segments of a pie) comprise the most popular methods of representing statistics. Although they involve a considerable amount of mathematical calculation, they are not difficult to draw. They can be used for a very wide variety of purposes and have the advantage of being able to present a striking visual impression. As is the case with many other types of graphs or charts and diagrams, they may be used—and often are used—in conjuction with a map in order to introduce the idea of location.

There are basically two types:
(i) Simple divided circle
(ii) Proportional divided circle and semi-circles.

1(i) Simple Divided Circle (Figs. 13(a), 13(b), 13(c), 13(d))

CONSTRUCTION

a. The circle may be of any convenient size; too small a circle must be avoided.
b. The circle is divided into segments which are proportional to the value of the individual components. Two methods may be used in calculating the size of the segments. In the first method, components are calculated as percentages of the total and, as 1% of the whole circle is equivalent to $3.6°$ (see Fig. 13(a)), segments can be drawn accordingly with a protractor (if a percentage protractor is available this will considerably reduce the

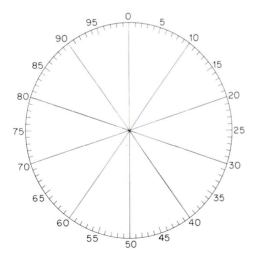

13(a) Simple divided circle or pie chart: a percentage chart amount of calculation necessary). Secondly, components may be expressed as a decimal or fraction of the total and the angle of the segment calculated as a decimal or fraction of 360°.

c. The largest component is usually placed to the right of '12 o'clock'.

d. Small segments often present a problem and are best grouped together around '9 o'clock' if possible—this will make the horizontal writing of names or figures easier. They should not be calculated last, as any error that accumulates (for example, from the width of the pencil lines) will be more noticeable on a small segment; such errors can be absorbed into the larger segments without unduly affecting the accuracy of the whole. To reduce errors to a minimum, angles should be measured on a cumulative basis: the second angle should be added to the first and the total thus obtained should be plotted; the third angle is then added to the preceding two angles and so on. By working clockwise from the vertical radius, this will minimise any error that may arise from the accumulated thickness of pencil lines.

e. All printing on the pie chart should be in block capitals and, as far as possible, horizontal. Do not write around the graph, turning it as you go. In the case of very small segments information can be added outside the chart and an arrow drawn from it to the relevant section (Fig. 13(c)).

f. It is often advantageous to mark either absolute or percentage values (as calculated above) on the relevant segment and to colour the segments distinctively; reference can then be made to a key. It is not advisable to try to print too much information on the chart.

g. There is normally no need to draw a concentric outer circle in which to print information; more especially this applies to proportional divided circles (see p. 24). Sometimes, however, further information can be given, for example concerning imports or exports, by adding an outer concentric arc, preferably as a dotted line, grouping together, between the relevant radii produced, imports from a particular group of countries or exports of a distinct character, for example mineral or agricultural exports (see Fig. 13(c)). In such cases segments that are similar in character must also be grouped together.

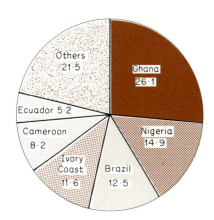

13(b) Simple divided circle: world production of cocoa beans by countries, 1968. Figures on the chart represent percentages of total world production of 1,249,000 tonnes. (Source: *Geographical Digest 1971*)

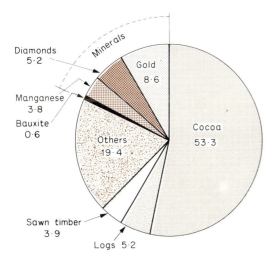

13(c) Simple divided circle: Ghana exports by value, 1967. Figures on the chart represent percentages of the total of N₵245,122,000 for exports.

h. Small segments should be coloured with a bright colour or shaded by some distinctive means. Generally, colouring is preferable to shading.

i. The number of segments will be dictated by the nature of the statistics; the greater the number of segments the more difficult the pie chart becomes to read. Seven or eight segments would seem to be the maximum number that can be represented easily, but much depends on the size of the circles—the larger the circle, the greater the number of segments possible.

GENERAL

a. A striking and effective visual impression can be given by a simple divided circle, especially if colours are used.

b. A simple divided circle, however, lacks the exactness of a bar graph, mainly because of the fact that reference cannot be made to a scale.

c. Accurate assessment of angles without measuring is not easy, but a quick evaluation can be achieved by using a percentage protractor. Statistical information relating to values or percentages may also be included on each segment.

d. Two or more simple divided circles of the same size may be used for the purposes of comparison, provided that the emphasis lies on the comparison of the components rather than on comparison of totals—for example, composition by race or tribe in different countries, exports for different years from the same country, annual allocation of national expenditure on services and such like, where, for the purposes of comparison, components are more important than totals (see Fig. 13(d)). But if consideration of totals is of equal or greater importance than of the components, then proportional divided circles should be used. A further discussion of this problem will be found in the next section.

1(ii) Proportional Divided Circles and Semi-circles (Figs. 14(a), 14(b))

CONSTRUCTION

a. When comparison of totals plays an essential part in representing statistics by divided circles it is necessary to construct two or more

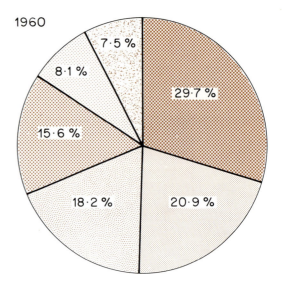

1960

Industry and handicrafts
Service trades, public administration, etc.
Agriculture etc.
Trade and commerce
Building and construction
Transport

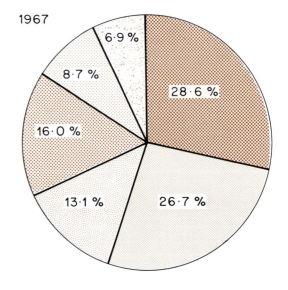

1967

13(d) Simple divided circle: percentage distribution of population by occupation for Denmark, 1960 and 1967. (Source: *Official Year Book*, Denmark 1970)

circles, each of which is proportional to the totals given.

b. To determine the radius of each circle it is necessary first to find the square roots of the totals to be represented by these circles, (π, being constant, need not enter into the calculations). Square root tables or a slide-rule will make the task of calculating easier.

c. For each circle components must be expressed as percentages, decimals or fractions of the whole, as in the simple pie chart.

d. The angles of each segment can be calculated as before.

e. Outer concentric circles containing information of an explanatory nature must not be drawn, as this would give the wrong impression of relative areas of the circles.

f. For purposes of comparison segments should be arranged in the same order in the circles.

g. Advice given in the construction of simple pie charts under sections (c), (d), (e) and (f) also applies to proportional divided circles.

h. Proportional pie charts may also be drawn as semi-circles adjacent to each other (see Fig. 14(a)) each semi-circle being proportional in area to the total quantity to be represented. The method of construction is essentially the same as that used for proportional pie charts. The difference lies in the fact that calculations must be based on half-circles and therefore, after calculating the radii and determining the angles of the segments, these angles must be halved (see Fig. 14(a)). Such a method is useful when limitations of space must be taken into consideration or when a more direct and obvious impression of comparison is intended. Although offering a compact and often striking impression, its uses are limited and it presents the same difficulties in interpretation as already discussed above; depending on the sizes of the semi-circles, seven or eight segments are usually sufficient.

GENERAL

a. A simple pie chart, drawn in isolation, presents little difficulty in interpretation, for the diameter, and therefore the area, is entirely arbitrary. It is only when proportional divided circles are drawn for the purposes of comparison that difficulties arise, when, to the problem of assessing areas as represented by

circles of different sizes, is added the difficulty in evaluating individual segments.

b. Comparing the relative areas of circles is always difficult and herein lies one of the weaknesses of the proportional divided circle. This may be overcome to some extent by indicating the total value or quantity of each circle and by including in the title the fact that the areas of the circles are proportional to the totals they represent.

c. It must be remembered that, as far as components are concerned, comparisons are not based on areas of segments (this would in fact be extremely difficult, if not impossible, and would seriously weaken the value of the chart), but on the angles at the centre—and therefore the percentage or fraction—that each segment occupies. To assist in the interpretation, the absolute or percentage value of each segment can be printed on the pie chart. Reference may then be made to a key for identification of the components.

d. When compared with the compound or comparative bar graph, a proportional divided circle gains from the fact that visually it is more pleasing and gives the impression of compactness and apparent simplicity in inter-

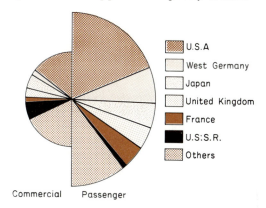

Key:
- U.S.A
- West Germany
- Japan
- United Kingdom
- France
- U.S.S.R.
- Others

Commercial   Passenger

'000's motor vehicles

	World total	U.S.A	W.Germany
Commercial	6 530	1896	241
Passenger	21,720	8 222	2 862

Japan	U.K.	France	U.S.S.R.
2 052	409	242	750
2 055	1 816	1 832	280

14(a) Proportional divided circle: semi-circular proportional divided circle or pie chart showing world production of motor vehicles (commercial and passenger) by countries (as percentages), 1968. Semi-circles are drawn proportional in area to the totals they represent.

2. DIVIDED RECTANGLES
(Figs. 15(a), 15(b))

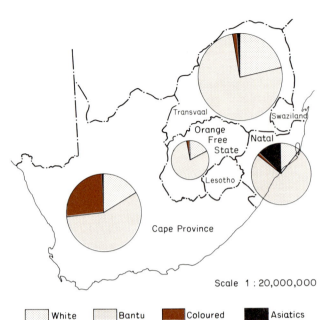

Scale 1 : 20,000,000

	White		Bantu		Coloured		Asiatics

	Total	Cape Province	Natal	Orange F.S.	Transvaal
Total	21,448,169	6,721,849	4,245,675	1,715,589	8,765,056
White	3,751,328	1,113,861	446,026	296,083	1,895,358
Bantu	15,057,952	3,824,175	3,212,909	1,383,233	6,637,645
Coloured	2,018,453	1,762,185	68,599	36,278	151,391
Asiatics	620,436	21,628	518,141	–	80,662

14(b) Proportional divided circles: racial composition of
Provinces of South Africa, 1970. Circles are drawn
proportional in area to the totals they represent.
(Source: *International Year Book and Statesmen's
Who's Who*, 1973)

pretation. But it may well be argued that this
advantage is only at the cost of a great deal
of mathematical calculation and that the final
result lacks the exactness in comparison which,
with the use of a scale, is possible with bar
graphs. Whereas the proportional pie chart
best sets the problem: 'How much, in terms
of percentage or fraction?', the compound and
comparative bar graph asks the question:
'How much, in terms of quantity or value?'

e. When used in conjunction with a map to give
some idea of location, proportional pie charts
present less of a problem than when using
compound or comparative bar graphs for this
purpose.

f. On balance, therefore, provided that the
reader appreciates the problems involved in
their interpretation and that assistance is given
in factual form on the graph itself, proportional divided circles may be regarded as a
useful form of statistical representation.

One of the most useful and most versatile methods
of statistical representation, but one which is
employed less frequently than one would expect,
is the divided rectangle, in which the total
quantity or value is represented by a rectangle
which is then sub-divided to indicate the constituent parts. Its lack of popular appeal may
perhaps be attributed to the fact that a circle is
generally more pleasing to the eye than a right-
angled figure; nevertheless, the divided rectangle has distinct advantages over the compound
bar graph, to which it is similar, and over the
pie chart, which it can usually replace.

Whereas the bar graph and the pie chart have
only one dimension that can be adjusted (the
length of the bar or the radius of the circle), the
divided rectangle has two dimensions that can be
varied. Because each division of the rectangle can
be subdivided, it is capable of much greater
flexibility and can be used to convey a great wealth
of information. If used as a purely diagrammatic
device without regard to the inclusion of all necessary and relevant scales, each sub-division can
be further sub-divided; one can devise, for
example, a rectangle drawn to represent the total
land area of, say, Australia, divided according
to the area of each state; each state can then be
sub-divided to show land use (for example,
arable, pasture, forest, waste); arable can then be
sub-divided into cereals and root crops; cereals
into wheat, barley, maize, oats and so on. But,
as this can only be achieved by confusing
horizontal and vertical scales, its value as an
accurate indication of relative values is limited,
although it could be pictorially useful.

As with bar graphs, two methods can be
employed: the simple divided rectangle and a
compound type.

CONSTRUCTION

2(i) Simple Divided Rectangle (Fig. 15(a))

a. A rectangle is drawn, the area of which is proportional to the quantity or value of all the
parts. As the vertical scale is usually chosen as
the 'constant', there is no difficulty in dealing
with totals (even prime numbers: for example,
673 units can be represented by a vertical scale
of 20 units and a horizontal scale of 33·65
units).

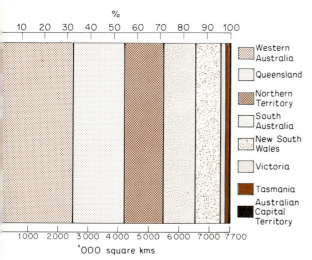

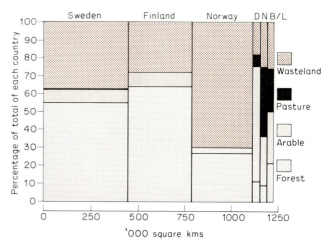

Australia: Area of states

Area of states	Square Kms
Western Australia	2,528,290
Queensland	1,727,970
Northern Territory	1,347,660
South Australia	984,630
New South Wales	801,640
Victoria	227,690
Tasmania	68,350
Australian Capital Territory	2,430
Total	**7,688,660**

15(a) Divided rectangle: area of states, Australia, 1971. The horizontal scale represents total area (sq km) and total area by percentage (top scale).

		Land use (percentage of total area, 1968)			
Country	Area ('000 sq kms)	Wasteland	Pasture	Arable	Forest
Sweden	450	37	1	7	55
Finland	337	28	–	8	64
Norway	324	70	–	3	27
Denmark	43	18	7	64	11
Netherlands	34	25	39	27	9
Belgium/Lux	34	26	24	29	21

15(b) Divided rectangle: land use in six European countries (Sweden, Finland, Norway, Denmark, Netherlands, Belgium/Luxembourg), 1968. The horizontal scale represents total area. (Source: *Geographical Digest*, 1971)

b. The rectangle is then divided into strips, each of which is of uniform height, variation in the values of the constituent parts being made possible by sub-divisions along the horizontal scale. For example, a total of 640 units could be represented by a rectangle 20×32 and then divided into its component parts (e.g. 390, 141, 75 and 34 would be represented as rectangles of 20×19.5, 20×7.05, 20×3.75 and 20×1.7).

c. If it is clearly borne in mind that the simple divided rectangle represents total value or area of the components and that the vertical scale remains constant, it can be seen that a vertical scale is not necessary (and should not be included). As this is essentially an areal method of representation, linear scales must be omitted. If the horizontal scale is drawn the same length as the horizontal axis, any unit of sub-division on the horizontal scale can be regarded as representing a column of constant height (i.e. the vertical constant) standing on that particular unit of sub-division. The single horizontal unit thus takes on an areal connotation and the whole horizontal scale can be

regarded as an areal scale. Sub-divisions of the horizontal scale are then made, not on a linear, but on an areal basis. No other scale is necessary. Information to this effect should be included on the chart—for example, 'the horizontal scale represents total (total areal) values'. In the same way, the horizontal scale may be regarded as 100% and then sub-divided (see Fig. 15(a)).

2(ii) Compound Divided Rectangle (Fig. 15(b))

a. Divisions of a rectangle may themselves be divided if the inclusion of further information is aimed at. In Fig. 15(b), for example, the total area of the six European countries is divided in proportion to their respective areas. Each country is sub-divided to show land use by area. As each division (i.e. each country) can be regarded as an entity, the vertical axis can then be scaled on a percentage basis and used to indicate the relative values of the sub-divisions.

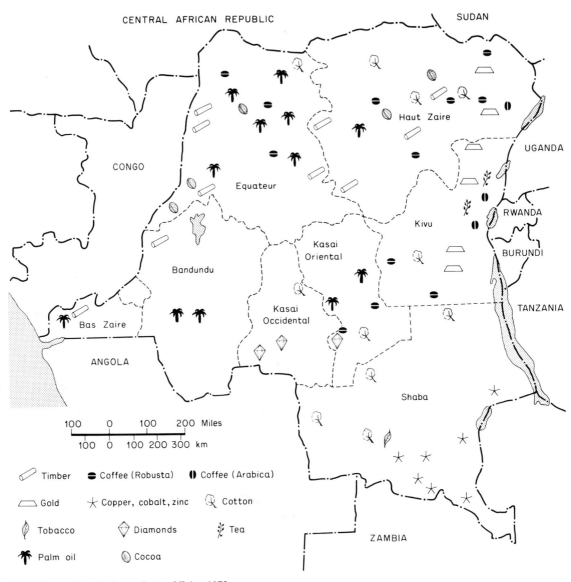

16(a) Repeated symbols: products of Zaire, 1972

GENERAL

a. Compared with a pie chart, the divided rectangle lacks the universal appeal of a circle, where each segment can be referred to a common point of origin and in which only the angle of each segment has to be assessed. A further disadvantage of the divided rectangle lies in the fact that it is of limited use for locational purposes.

b. The main advantage of the divided rectangle lies in its simplicity of construction; unlike the divided circle it requires no involved or tedious mathematical calculations. In its simple form it resembles a compound bar graph but it must be remembered that both its dimensions have significance. The divided rectangle has the added advantage that it can convey much more statistical information than either a compound bar graph or a divided circle. Unlike the latter, reference may be made to a visible scale (both horizontal and vertical, if desired), while the task of assessing the comparative areas of two rectangles of different sizes is no more difficult (and probably less) than assessing the relative areas of two circles of different diameters. The fact, too, that it can initially be constructed on graph paper is a further advantage.

c. This, then, is a method of statistical representation which is relatively easy to construct, has definite advantages and could certainly be used more widely.

3. REPEATED SYMBOLS (Figs. 16(a), 16(b))

Statistical information can be represented on a map by the repetition of one symbol of uniform size or character, or by a variety of symbols. The idea of location is expressed by placing the symbol as nearly as possible in its correct place on the map.

Two types of repeated symbols can be distinguished: the non-quantitative type (see Fig. 16(a)), which is basically pictorial or descriptive (for example, crops by their initial letter—C for cotton, F for fruit, R for rubber—illustrations of plants for crops, drawings of cattle for ranching, pictures of trees for forests, a range of symbols for minerals and so on) and, secondly, the quantitative type, which, by grouping the symbols together, seeks to represent totals (see Fig. 16(b)).

10 0 20 40 60 km

➤ About 30,000 tons of landed fish

Fish landings, 1968	
	'000's metric tons
Esbjerg	551
Skagen	355
Hirtshals	282
Thyboron	125
Hvide Sande	40
Strandby	32
Frederikshavn	15
Grena	10
Total	1 410

16(b) Repeated symbols: Denmark's fishing and fishing ports

Repeated symbols form one of the simplest methods of recording statistical information and can still be frequently seen on maps dealing with agricultural products, minerals, 'economic' development, maps and guides produced for tourists, for advertisement purposes and the like.

The non-quantitative type of repeated symbol is essentially a descriptive device giving a visual impression which reminds one of the lines of Jonathan Swift:

So geographers, in Afric maps,
With savage pictures fill their gaps ;
And o'er unhabitable downs
Place elephants for want of towns.

Oil wells, for example, may be marked by a symbol resembling a derrick, coal mines by a lump of coal or pit-head workings, but to mark an oil-field or a coal-field or any aspects of their relative importance by this method is out of the question. Nor can the placing of the symbol to represent, for instance, cocoa or coffee production coincide with any degree of accuracy with the actual area of production. Although an attempt is sometimes made to show relative values by adjusting the size of the symbol (usually arbitrarily), there is no gain in accuracy. Generally speaking, therefore, non-quantitative symbols aim at a pictorial representation and a visual impression which too often results in the subordination or even suppression of locational and statistical accuracy.

The same type of symbol may also be used on a quantitative basis to represent statistical information, such as quantity or value. Symbols are grouped together, the total number of symbols being proportional to the total quantity to be represented. One cocoa bean, for example, may represent 10,000 tonnes of cocoa, one cotton bale 10,000 tonnes of cotton, one man one million people. Geometrical shapes, such as squares, cubes or circles, may also be used; different colours may be employed to indicate different categories or types. It is then possible to count the number of symbols and, by reference to the key, calculate the total value. But it suffers from many disadvantages: it is difficult to represent a wide range of values, the drawing of the symbols is often a tedious and difficult task, the general impression is one of over-simplification, the group of symbols rarely coincides with the area of production, while the symbols themselves could quite easily—and more effectively—be replaced by a figure printed on the map, which would give the same information in a more readily assimilable form.

4(i) PROPORTIONAL CIRCLES
(Figs. 17(a), 17(b), 18(a), 18(b))

Mention has already been made (under the section dealing with proportional divided circles) of the use of the circle as an areal symbol and, indeed, this section can be regarded as an extension of the same idea, its purpose being mainly to describe the various methods that can be employed in the construction of proportional circles.

METHOD 1 (see Fig. 17(a), i)

Although it may involve a considerable amount of mathematical calculation, the use of square roots is a straightforward method of obtaining the radii of the circles required. Having obtained the square root, the choice of a scale is a matter of discretion. The method, however, suffers from the disadvantage that, if a large range of numbers is to be represented, the largest circle may be too large and the smallest too small; for example, when dealing with a range of numbers extending from 10,000 to 1,000,000 (such as population of towns), if the smallest value (10,000) is represented by a circle of 2 mm radius, the largest value would require (on the same scale) a circle of 2 cm radius.

To overcome these two difficulties—the calculations involved and the necessity to limit the

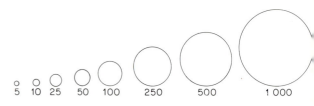

(i) Method 1

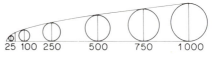

(ii) Method 2

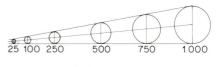

(iii) Method 3

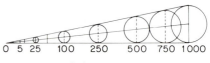

(iv) Method 4

17(a) Four methods of constructing proportional circles

range of size of circles—it is often advantageous to construct a scale from which radii of circles, proportional to each other, can be measured directly with dividers, more especially if it is necessary to draw a large number of proportional circles.

METHOD 2 (see Fig. 17(a), ii)

A line of any convenient length is drawn, which is then evenly graduated from zero on the left to the maximum value to be represented on the right. Calculate the square roots of several values (for example, in Fig. 17(a) ii, 25,000, 100,000, 250,000, etc. were chosen) and from the corresponding point on the horizontal scale already drawn draw perpendiculars proportional in length to these square roots. The ends of these perpendiculars are then joined by a smooth curve (notice that it will not be a straight line). Perpendiculars of intermediate values can then be measured by dividers or ruler: these measurements will represent the radii of the required circles.

METHOD 3 (see Fig. 17(a), iii)

A variation of the previous method, which is easier to draw but less mathematically accurate, is as follows. Having first decided on the minimum and maximum size of circles to be used to represent the range of values, draw a linear scale as before (i.e. evenly graduated). At one end draw a perpendicular equal in length to the radius of the smallest circle to be used and at the other end a perpendicular of the same length as the radius of the largest circle. The ends of these perpendiculars are then joined by a straight line. Intermediate values can then be plotted along the horizontal scale: the length of the perpendiculars from these points will be the radii of the required circles.

METHOD 4 (see Fig. 17(a), iv)

This method employs a linear scale which is graduated from zero (usually on the left) in such a way that the spaces between the subdivisions are proportional to the square roots of the values. At the other end draw a perpendicular, the length of which is proportional to the radius of the largest circle to be used; a straight line is then drawn between the end of the perpendicular and the zero mark. The length of the radius of any desired

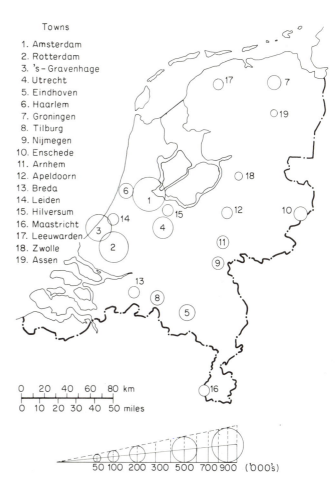

Towns

1. Amsterdam
2. Rotterdam
3. 's-Gravenhage
4. Utrecht
5. Eindhoven
6. Haarlem
7. Groningen
8. Tilburg
9. Nijmegen
10. Enschede
11. Arnhem
12. Apeldoorn
13. Breda
14. Leiden
15. Hilversum
16. Maastricht
17. Leeuwarden
18. Zwolle
19. Assen

0   20   40   60   80 km
0   10 20 30 40 50 miles

50 100 200 300 500 700 900 ('000's)

17(b) Proportional circles: population of major towns of the Netherlands, 1968

circle can then be measured by drawing perpendiculars from the appropriate point on the linear scale. The radius of a circle, for example, to represent 400,000 (using Fig. 17(a)iv) can be calculated as follows. The complete horizontal scale represents 1,000,000, i.e. $1,000^2$. The square root of 400,000 is 632·5. Therefore, along the horizontal scale mark a point $\frac{632 \cdot 5}{1,000}$ of the distance from the zero point. As the actual length of the horizontal scale is 10 cm, this is equivalent to 6·32 cm (N.B. Fig. 17(a) is drawn to half-scale). From this point draw a perpendicular to meet the oblique line—this vertical distance will be the radius of the circle required. This method is used in Fig. 17(b).

GENERAL

a. The inclusion of a key is essential. Using Method 1, a word statement only is sufficient—

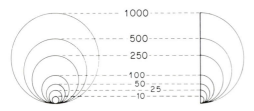

18(a) Circles and semi-circles used in drawing key for proportional circles

18(b) Methods used in drawing vertical scale for proportional circles

for example, 'areas of the circles are proportional to the quantities or values represented'. In the case of the other methods, the key may take the form of Figs. 17(a)ii, 17(a)iii, or 17(a)iv—as used in the construction. In all cases the key may be drawn as circles (semi-circles are sufficient) inside each other, standing on a common base (Fig. 18(a)), or simply in the form of the vertical scale only (Fig. 18(b)).

b. Do not confuse proportional circles with the graduated range of symbols (see page 36).

c. Proportional circles are often used in conjunction with a dot map (see page 42), but remember that there is no mathematical relationship between the size of the dot and the area of any of the circles.

d. As with the pie chart, proportional circles may be sub-divided to show components; they may also be superimposed on a base map for locational purposes.

e. One serious disadvantage is the fact that it is difficult to assess relative areas and hence values of different sized circles (see page 1).

f. The use of method 2, 3 or 4 (page 31) in constructing the circle has the decided advantage that a complete range of values, however large, can be represented; this is possible because the size of the smallest and the largest circles to be used is first fixed.

4(ii) PROPORTIONAL SQUARES
(Figs. 19(a), 19(b))

Proportional squares may be used in the same way as proportional circles. The area of the square is proportional to the quantity it represents and thus the length of the side of the square is directly related to the square root of the number to be represented. This method of representation is useful when there is a wide range of numbers to deal with, as the use of square roots reduces the range and makes representation more manageable.

CONSTRUCTION

a. Calculate the square root of the total to be represented; this, drawn to scale, will be the length of the side of the square. Square root tables will be of great assistance.

b. The key must be clearly shown. Two (basically similar) methods are possible; both are shown in Fig. 19(b).

c. If it is necessary to show part of the whole (i.e. to show a sub-division) the same process is followed: the square root of the part will give the length of its side.

GENERAL

a. Proportional squares may be employed without reference to a map, in which case they are usually drawn in a row (see Fig. 19(a)). They are more often, however, used locationally on a base map (see Fig. 19(b)) for purposes of comparison.

b. Although squares are more difficult to draw, their relative areas are probably more easy to assess than circles. Graph paper will be found very useful in constructing the squares; a tracing can then be made on to plain paper.

c. In addition to showing comparative amounts, by their area or size, proportional squares can also be coloured or shaded (without sub-dividing) to indicate the character of the individual totals. The total amount of sugar produced in selected countries, for example, may be shown by the area of the square itself, while colouring or shading can indicate the average yield per acre/hectare for that country, or the type of sugar produced and so on. This, of course, necessitates the inclusion of another key.

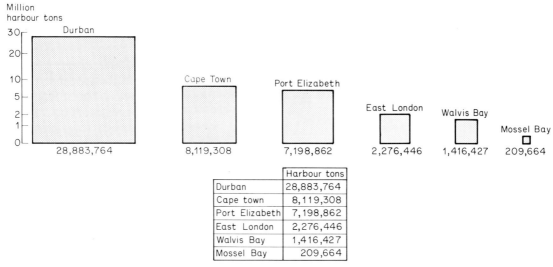

	Harbour tons
Durban	28,883,764
Cape town	8,119,308
Port Elizabeth	7,198,862
East London	2,276,446
Walvis Bay	1,416,427
Mossel Bay	209,664

19(a) Proportional squares: tonnage of goods handled at South African ports, 1970

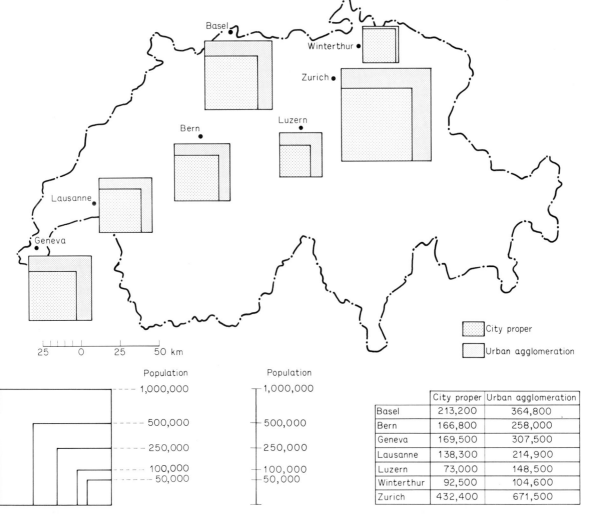

	City proper	Urban agglomeration
Basel	213,200	364,800
Bern	166,800	258,000
Geneva	169,500	307,500
Lausanne	138,300	214,900
Luzern	73,000	148,500
Winterthur	92,500	104,600
Zurich	432,400	671,500

19(b) Proportional squares: population of major cities and urban agglomerations of Switzerland, 1969. (Source: *U.N. Demographic Year Book*, 1969)

d. Squares may also be superimposed on each other, as in Fig. 19(b), to show increase or decrease, exports and imports, parts of a whole and so on.

e. Proportional squares are generally sited on the map in such a way that the south-west corner of the square locates the position of the town to which it refers. But, as with other proportional symbols, overlapping presents problems in drawing and in interpretation, although it gives an immediate impression of concentration or congestion. As a general rule, however, where overlapping is unavoidable, proportional circles are preferable.

f. Assessing small differences in total values represented by proportional squares is more difficult than in the case of a bar graph. In Fig. 19(a), for instance, the difference in tonnage of goods handled at Cape Town and Port Elizabeth amounts to almost 1 million tons. There is usually room, however, either on the square itself or below it, to indicate the value, if it is considered important that the figure should be readily available.

4(iii) PROPORTIONAL CUBES
(Figs. 20(a), 20(b))

Like proportional squares, proportional cubes can be drawn independently of a base-map and they are therefore included here. They are usually, however, given locational significance by being located on a base-map to represent quantitative distribution or production. (Fig. 20(b)).

Their main advantage lies in the fact that they introduce a third dimension and, just as areal methods of statistical representation (such as divided rectangles or proportional circles) have an advantage in certain circumstances over linear methods, so proportional cubes gain from the fact that they are capable of representing statistics which have an even greater range of values. In the same way as the length of the side of a proportional square bears direct relationship to the square root of the quantity to be represented, so the side of the cube is directly related to the cube root of the quantity. Thus, whereas a square with sides 5 mm long can be used to represent 25 units of production, a cube of the same measurements will represent 125 units.

But in this ability to 'compress' a wide range of values lies the inherent weakness of the proportional cube (see also Introduction, page 1).

CONSTRUCTION

a. The cube root of the quantity to be represented must first be calculated; to avoid lengthy calculations cube root tables are necessary.

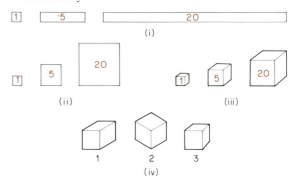

20(a) Comparison of (i) linear method of representation, (ii) proportional squares and (iii) proportional cubes. (iv) shows three types of cube: 1 and 2 are isometric, 3 is perspective. In the first three methods the figures indicate units.

b. If the cubes are to be drawn independently of a base-map, they should, for purposes of comparison, stand individually on a straight line (see Fig. 19(a)).

c. The cube can be drawn in several ways (see Fig. 20(a)): either isometrically, in which all sides are of equal length, or 'perspectively', in which the sides are one-half to three-quarters the length of the front. On any one map all cubes must, of course, be drawn to the same 'pattern'.

d. A key should be provided; an illustration is given in Fig. 20(b). Notice that cube-root values are plotted on the base-line.

e. Because of the unavoidable difficulty in evaluating quantities represented by cubes of different sizes, it will help to print the relevant quantity on the face of the cube.

GENERAL

As a symbol, a cube is easier to draw than a sphere and assessment of the relative volumes of two cubes of different sizes is probably less difficult than the relative volumes of two different spheres. But, as in the case of divided rectangles, a cube is, generally speaking, less pleasing visually than a sphere, unless the commodities concerned are themselves in cuboid form—for example, containerised traffic at a port,

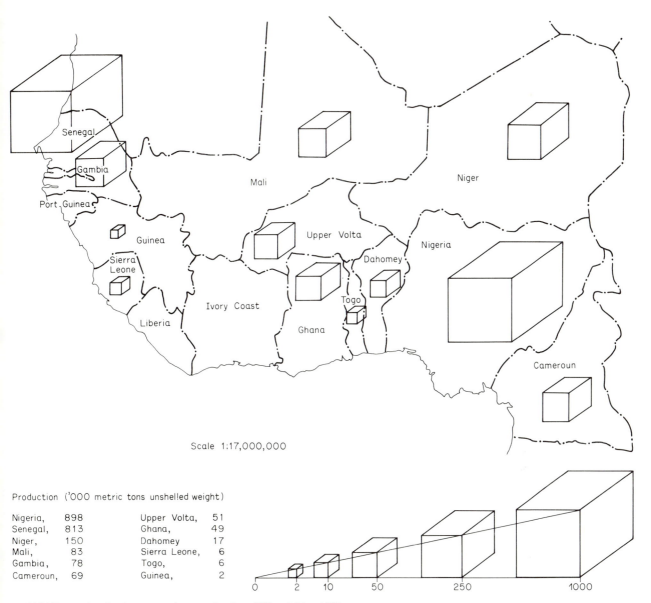

Scale 1:17,000,000

Production ('000 metric tons unshelled weight)

Nigeria,	898	Upper Volta,	51
Senegal,	813	Ghana,	49
Niger,	150	Dahomey,	17
Mali,	83	Sierra Leone,	6
Gambia,	78	Togo,	6
Cameroun,	69	Guinea,	2

20(b) Proportional cubes: groundnut production of West Africa, 1970

exports of tea in chests, or bales of cotton. Another disadvantage is the fact that the cube cannot be sub-divided to show components. Moreover, the comparison of quantities as represented by cubes is extremely difficult, unless statistical information is printed on the cube. Indeed, it seems very doubtful if proportional cubes have any great advantage over the printing of actual statistics on the map in their relevant places. Its use, therefore, is restricted, its limitations outweighing its advantages as a means of representation.

4(iv) PROPORTIONAL SPHERES (Fig. 21)

The proportional sphere is so similar in concept to the proportional cube that a brief description will suffice. It serves the same purpose as the cube in that the introduction of a three-dimensional figure allows a very wide range of values to be represented, the volume of the sphere being proportional to the quantity involved. Although probably more pleasing visually, it suffers from much the same disadvantages as the cube.

CONSTRUCTION

a. The radius of the sphere is determined by calculating the cube root of the quantity to be represented. (The volume of a sphere is $\frac{4}{3}\pi r^3$; when calculating the radii of more than one sphere, $\frac{4}{3}\pi$, being a constant, can be omitted from calculations and the cube root only taken into consideration).

b. The sphere is then drawn in its correct location on the map. Proportional spheres are not normally used except on a locational basis.

c. The actual drawing of the sphere to give a three-dimensional impression is by no means easy. Several methods can be used; some examples are given in Fig. 21.

d. As an alternative method for the key, a selection of various sizes of spheres, with information relating to the value of each, can be drawn.

21   Four methods of drawing proportional spheres

GENERAL

This method of statistical representation should (like cubes) be reserved for occasions when the range of statistics is so great that other methods are impracticable. It is sometimes used in conjunction with dots or proportional circles to avoid the overcrowding that would otherwise result when attempting to represent the distribution of population of a densely populated area within a larger, less densely populated region.

Apart from the difficulty in drawing and the calculations involved if cube root tables are not available, its value to the general reader is seriously diminished by the fact that it is almost impossible to assess the relative volumes, and hence values, of different spheres. It is difficult to appreciate, for example, the fact that a sphere of 4 mm radius represents a quantity eight times as great as a sphere of 2 mm radius and a sphere of 5 mm radius almost sixteen times as much as one of 2 mm.

Nor can a sphere be sub-divided to show constituent parts, while the area of the map covered by the sphere bears little relation to the area to which the statistics refer. The printing of statistical information on a sphere also presents a much greater problem than in the case of a cube and difficulty may be experienced in trying to draw a sphere small enough to represent the smallest value in a range of statistics.

5. GRADUATED RANGE OF SYMBOLS (Figs. 22(a), 22(b))

Apart from methods of representing quantitative distribution that have already been described, it is also possible to devise a range of symbols which do not depend on a close mathematical relationship to the quantities to be represented. Such symbols are usually drawn in the form of circles or squares, each symbol representing a definite range of values and indicating, either by an increase in size or by its appearance, increases in quantities (see Fig. 22(b)). This method of statistical representation is often used in connexion with population maps and can be found frequently in atlas maps depicting location and relative sizes of towns and cities.

CONSTRUCTION

a. First decide on the number of symbols to be used and the range of values that each will represent—this will be determined by the total range of the statistics. For this purpose the construction of a dispersion graph (see page 16) will be helpful. All values from zero to the maximum should be included; do not have gaps in the scale. In the case of population maps, critical values could, for example, be 10,000, 50,000, 100,000, 250,000 and 500,000, or, 100,000, 250,000, 500,000 and 1,000,000.

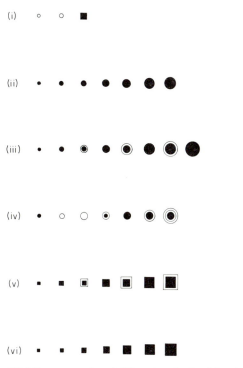

(i)

(ii)

(iii)

(iv)

(v)

(vi)

22(a) Some examples of different methods of drawing graduated ranges of symbols.

Although much will depend on the scale of the map, about eight or nine symbols will probably be sufficient.

b. Symbols can range from a small circle or dot to a square; usually, the square is used to represent the higher values. If it is necessary to increase the range of symbols, intermediate values can be represented by the addition of an outer circle or square to the previous symbol. In any case, the range of symbols used should give the appearance of increasing values. Examples are given in Fig. 22(a).

c. If the number of symbols to be used is small, a graduated range of circles only may be used; these should not, however, be confused with proportional circles (see page 30).

d. Both the key and the symbols must be drawn clearly and accurately.

e. This method can also be used for other purposes; for example, the distribution of factories and the number of people employed in each, of H.E.P. stations and their generating capacities, of towns in relation to specified services, and so on. The reader must remember, however, that other methods are available which are often more suitable and more effective.

GENERAL

The main advantage in this type of statistical representation is that it avoids mathematical calculations such as square or cube roots and it can be of use where other methods, such as dots, may result in overcrowding. Moreover, the number of symbols used can be adjusted to cover a small or large range of values; ingenuity can be used in devising an extended range of symbols, but, because constant reference must be made to a key, all symbols must be clear and easily recognisable. As is the case with shading (choropleth) maps (see page 45), it is advisable to prepare beforehand a range of symbols which can be readily used when necessary.

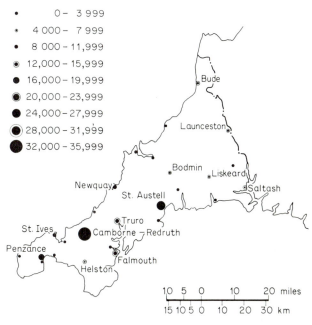

•	0 – 3 999
⊙	4 000 – 7 999
•	8 000 – 11,999
◉	12,000 – 15,999
●	16,000 – 19,999
◉	20,000 – 23,999
●	24,000 – 27,999
◉	28,000 – 31,999
●	32,000 – 35,999

22(b) Graduated range of symbols: population of major towns in Cornwall

Because of its ability to represent large values by a simple, relatively small symbol, this method can often be used in conjunction with other types of statistical maps to provide additional information on the same map. Graduated symbols, for example, are sometimes superimposed on world or continental maps that depict density of population, but, because it may result in ambiguity, mention should be made in the key to the effect that the centres of population represented by symbols are taken into account when calculating densities.

Although easy to draw, this method, however, has the serious disadvantage that no direct mathematical relationship exists between either the size or the character of the symbol used to represent a particular quantity and the quantity itself: the choice of symbols is entirely arbitrary. Because each symbol has a distinct and different value from any other symbol, it cannot be used in the place of a dot map, nor can it be employed to show density. It is, in effect, a pictorial or descriptive method of statistical representation, resting on no mathematical basis.

6. WIND ROSES

As its name implies, the principal use of the wind rose is to show diagrammatically the average frequency and direction of the wind at a given place; like many other types of statistical representation, it may be given locational significance by being drawn in its correct place on a map. It is essentially a linear method of representation, the direction and length of the line or column representing the direction and frequency respectively of the wind. In the case of the compound wind rose the width of the column may also be adjusted to indicate wind speed.

6(i) Simple Wind Rose (Fig. 23(a))

CONSTRUCTION

a. The centre of the wind rose is usually a circle of any convenient size. Instead of a circle, an octagon may be used, each side representing one point of the compass.

b. The number of days of calm, reckoned either monthly or annually, are represented separately. They are usually indicated by a figure which is printed inside the circle (see Fig. 23(a)) or, sometimes, immediately under the wind rose. The circle itself can be used to indicate the number of days of calm by making it proportional in area to the total number of such days; in this case, an explanatory note and a key must be added. If percentage values (see paragraph 'd' below) are used in calculating the length of the columns or 'arms', then calms also should be represented by a percentage scale (see paragraph 'e' below), printed inside the circle or underneath it.

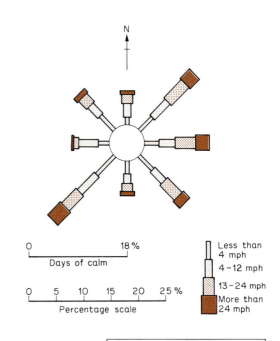

Direction of wind	N	NE	E	SE	S	SW	W	NW	Calm
Number of days	3	2	1	1	3	7	7	5	2

23(a) Simple monthly wind roses

		Wind direction by percentage							
Wind speed		N	NE	E	SE	S	SW	W	NW
Less than 4 mph		1·3	3·2	2·6	3·4	2·2	4·6	2·0	3·6
4 – 12 mph		2·4	4·0	3·1	2·8	1·7	4·4	3·5	2·5
13 – 24 mph		1·6	4·5	3·6	3·9	1·4	3·7	1·0	2·1
More than 24 mph		0·8	2·5	2·6	1·5	0·8	3·4	0·3	1·0
Totals		6·1	14·2	11·9	11·6	6·1	16·1	6·8	9·2

Calms: 18·0

23(b) Wind rose: mean annual percentage frequency of wind direction and wind speeds

c. Eight points of the compass are usually sufficient, but sixteen points can be used if greater detail is required—this, however, is rarely necessary.

d. The length of the columns or 'arms' can be drawn proportional to the actual number of days; for example, a north-east wind for ten days in a month would be represented on a monthly wind rose by a column 10 units long, and so on. Alternatively, frequency of the wind from each point of the compass may be calculated as a percentage of the whole and a percentage scale added; this method is much more usual in the case of annual frequency diagrams.

e. The drawing of a percentage scale often presents difficulties, but, if properly understood, should not do so. The total number of days (in the month or year) is regarded as 100%—this will include days of calm. A line of convenient length is chosen to represent 100%, percentage frequency of wind is calculated for each point of the compass and columns of correct percentage lengths are then drawn. Statistics, for example, for a station in the month of June might read: north, 0 days; north-east, 2; east, 4; south-east, 7; south, 6; south-west, 4; west, 3; north-west, 1; calm, 3 days. These would be equivalent to 0%, $6\frac{2}{3}\%$, $13\frac{1}{3}\%$, $23\frac{1}{3}\%$, 20%, $13\frac{1}{3}\%$, 10%, $3\frac{1}{3}\%$ and, for calms, 10%. It is not necessary to draw on the wind rose the complete percentage scale (i.e. for 100%) (see Fig. 23(b)). Usually, one-quarter or one-third only of the length need be drawn, sub-divided up to 25% or $33\frac{1}{3}\%$ respectively. Any other convenient portion may be drawn, but it should not, in any case, be shorter than the length of the longest column.

f. Columns can be replaced by straight lines of correct relative lengths, the number of days being represented by an equivalent number of small bars drawn at right angles to the line and equally spaced along it (see Fig. 23(a)).

CONSTRUCTION

a. The basic method of construction is identical to the simple wind rose, whether actual or percentage values are used.

b. The speed of the wind is then indicated by varying the width of the column—an increase in width signifies an increase in wind speed. The divisions usually chosen are: less-than-4 mph, 4–12 mph, 13–24 mph and more than 24 mph. These divisions correspond with the Beaufort scales of 1, 2–3, 4–5 and 6 or over (calm is force 0, less than 1 mph). Wind speed may also be recorded in knots or in metres per second (see Appendix, page 59, for metric conversion tables).

c. Winds with least velocity are represented nearest the centre.

d. Colouring or shading of the different sections of the column will make interpretation easier. A key must be added.

GENERAL

As a method of representing statistics, the wind rose is ideally suited to its purpose, as it shows instantly and accurately the average direction and frequency of winds over a period of time and can also give a visual impression of wind speeds from which comparisons can be made.

Any defects as a method, however, are of less consequence than the limitations in its interpretation. The 'mean annual frequency' rose, for example, obscures seasonal patterns and movements of wind, although this can be overcome to a large extent by constructing wind roses for each month. Further detail, if required, can be supplied by sample daily wind roses, if it is desired to show the incidence of, for instance, land and sea breezes or other daily variations.

6(ii) Compound Wind Rose (Fig. 23(b))

A modification of the simple wind rose may be employed to indicate not only direction and frequency but also speed of the wind.

FOUR
Statistical Maps

1. DOT MAPS (Figs. 24(a), 24(b), 24(c))

One of the simplest and most widely used types of distribution map is the dot map. Its simplicity is attributable to the fact that basically it is a combination of two straightforward ideas: the repeated symbol, in this case a dot of uniform size (see page 29), and a dispersion map on which dots are placed locationally. But before describing its construction and uses it is essential to make a clear distinction between two terms that are often confused: distribution and density. A distribution map is essentially the representation of absolute or actual quantities on a map; the dots used on a dot map are quantitative symbols, each having a specific and fixed value (Fig. 24(a), 24(b), 24(c)). It is thus possible (although not always practicable) to count the number of dots on this type of map and then, multiplying by the dot value, to calculate with some accuracy the total value. A density map, on the other hand (see page 45), is concerned with averages, percentages or ratios (for example, number of people or number of sheep or yield of rice per hectare)—in other words a relationship between quantity and area, pre-supposing a uniform density throughout a given area. Obviously, if a dot (distribution) map is drawn accurately, it will give an overall impression of comparative densities, but this is not the basis on which it was constructed. One of the great advantages of the dot map is its wide variety of uses. Distribution of 'commodities' of an almost limitless range (including volume, area, value, weight and number, can be represented and it is not surprising that this method is so frequently used.

Its construction, nevertheless, presents many problems both in the preparation and drawing stages.

CONSTRUCTION

Attention must first be given to four problems that will present themselves: (a) the numerical value of the dot, (b) its size, (c) its location and (d) how to draw dots of uniform size.

a. Dot-value

It is first necessary to decide what one dot shall stand for (e.g. how many people? how many tonnes? what value?) This will, of course, determine the number of dots to be placed on the map. Too many dots (i.e. when the dot-value is too low) will produce a map that is overcrowded, especially in areas of greater concentration; too few (if the dot-value is too high) will give an equally wrong impression. In deciding on the dot-value, attention must be given to the range of figures to be represented, as this will directly affect both the value and the number of dots to be drawn. If possible, a quick 'trial' map should be prepared; this need not be completely accurate, but it will repay the time and labour involved, as a great deal of the success or otherwise of the map will depend on the decision concerning dot-value.

b. Dot-size

The size of the dot to be drawn (i.e. its diameter) must also be considered, but, clearly, this cannot be decided in isolation—dot-value and dot-location must also be taken into account. For obvious reasons extremes of sizes must be avoided; the number and size of the dots must be such that they convey a clear visual impression of the differences in distribution, contrasting, as far as possible, areas of greater concentration with areas of a more sparse or scattered nature.

c. Dot-location

Two methods are possible in placing the dots. The first is of limited value, although it is sometimes necessary to adopt this method in the absence of sufficient relevant information affecting the distribution. Having calculated the number of dots to be used, these are then distributed evenly over the area concerned (the 'statistical unit') (see Fig. 24(a)). This method has obvious disadvantages: it conveys little information except the total quantity (i.e. the number and value of the dots) and gives no indication concerning the true distribution. It does, however, give a visual impression of comparative densities and it is possible to construct a key indicating 'sample densities', but this is not its purpose and

is better represented by a shading (choropleth) map (see page 45).

The second method takes into account not only the quantities to be represented but also attempts to place the dots in their correct position on the map to give a more accurate impression of the distribution as in Figs. 24(b) and 24(c). This can best be done by making use of first-hand knowledge of the area, but, in the absence of this, a reasonably accurate map can be constructed by using information gathered from other maps of the area, for example, maps depicting relief and drainage, geology, soils, vegetation, rainfall, land use, communications, water supply and settlements. By analysing and collating such maps it is possible to concentrate dots in certain areas and to avoid others. This will not necessarily produce a completely accurate map but will, at least, give a more correct impression than an even distribution of dots would achieve.

Difficulty will no doubt arise in dealing with areas of denser concentration (see also section (i) below). In all cases, it is advisable to tackle these areas first. Calculate the number of dots that will be required in any particular 'problem' areas and plot these first; distribution of the remainder of the dots over the rest of the map can then be undertaken.

d. Drawing the dot

The actual drawing of the dots often presents a very real problem to the 'amateur'. Preparation should first be made by marking on the map the position of all dots, very lightly, in pencil. The subsequent drawing of dots of the correct size, circular and uniform in character, cannot successfully be done with an ordinary pen or pencil. If available, dotting pens with round, flat tips of various sizes, or specially constructed 'Uno' pens may be used. Punches, too, may be employed; either the small confetti-like discs can be glued on to the map or punch-holes of the correct size may be made on the map and then the whole map mounted on black paper. It is possible, also, to use a stencil, made from celluloid or plastic, in which holes of varying sizes have been cut; although this is not suitable if ink is to be employed, it can be used with coloured pencils.

Probably the most satisfactory method for the non-professional is to use fibre- or nylon-tipped pens which are now readily and cheaply available (felt-tipped pens are not firm enough for this purpose). If used carefully, reasonable success can be achieved in making circular dots of uniform

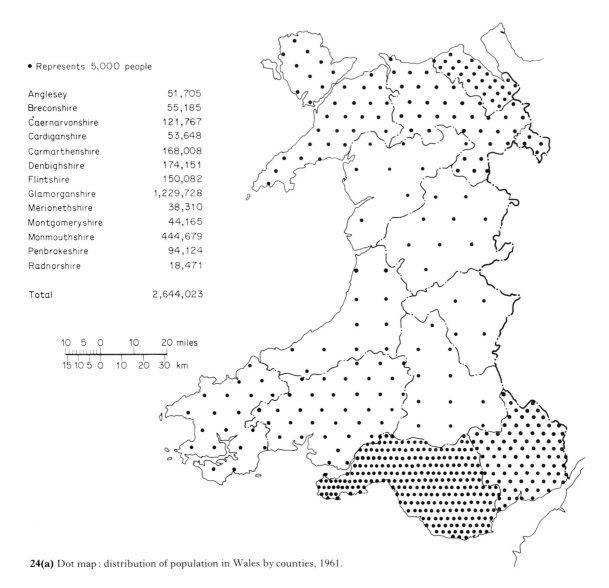

● Represents 5,000 people

Anglesey	51,705
Breconshire	55,185
Caernarvonshire	121,767
Cardiganshire	53,648
Carmarthenshire	168,008
Denbighshire	174,151
Flintshire	150,082
Glamorganshire	1,229,728
Merionethshire	38,310
Montgomeryshire	44,165
Monmouthshire	444,679
Penbrokeshire	94,124
Radnorshire	18,471
Total	2,644,023

24(a) Dot map: distribution of population in Wales by counties, 1961.

size. These pens can be sharpened or blunted on fine sandpaper to produce smaller or larger dots. Provided that the paper is not absorbent and that the pen is held vertically with a consistently even pressure, this method will produce a good result. It has the added advantage that fibre or nylon-tipped pens can be obtained in a wide range of different colours. To avoid a 'tail' attached to the dot it is essential to lift the pen vertically and slowly from the paper; it may also help to slide the fingers lightly downwards over the pen, at the same time slightly releasing the grip, before lifting the pen.

i. Drawing dots in 'high-value' areas always presents a problem. If the dot map is to retain its character and its characteristic of being an essentially distributional map, representing absolute values by dots which have a specific value, it should not be combined with other methods of statistical representation on the same map. Proportional circles, squares and, more especially, spheres cannot be evaluated on the same basis as the dot, the former being mainly areal symbols, the latter numerical. Similarly, if an inset, on an enlarged scale, of a 'high-value' area is included, the dots should remain of the same size and a note concerning the scale of the inset added. As it is not the intention of the map that dots should be counted, the fact that in some places dots merge and form almost a solid black area does not necessarily detract from its value. Solutions to the problem may lie in an adjustment of the dot value (too small a dot value may have been chosen) or of the size of the dot, a larger scale base-map or the construction of two maps, one by dots and the other wholly by symbols such as proportional circles which can be varied in size according to the quantity they represent.

ii. The smaller the statistical area or unit, the more accurately can dots be placed.

iii. The dot value should be kept as low as

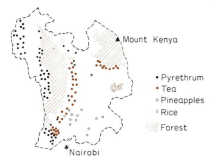

1 dot represents 200 hectares

24(b) Dot map: distribution of pyrethrum, tea, pineapples and rice in Central Province of Kenya

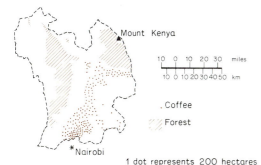

1 dot represents 200 hectares

24(c) Dot map: distribution of coffee in Central Province of Kenya

possible, provided that the map does not become so overcrowded as to impair its value as a distribution map.

iv. Boundaries of statistical units (parish, district, county and so on) should not be shown on the finished map, although such boundaries may be necessary in its construction (see Fig. 24(a)). Dots, therefore, should be placed close to or even on the boundaries so that, when the map is complete, former boundary lines are not obvious, like fire-breaks in a forest.

v. The key should include reference to the value of the dot and, apart from the title, a scale.

vi. The distribution of two or more commodities may be shown on the same map by dots

of different colours (but the same size). Bright colours should be used, to make identification easier, but it is a mistake to try to show too many commodities.

vii. The dot-value may also be calculated on a percentage basis. In this case, statistical units are calculated as a percentage of the whole. In Fig. 24(a), for example, the population of Glamorganshire (see table) represents approximately 46%, Carmarthenshire 6% and Anglesey 2% respectively of the total population of Wales. Assuming that one dot represents 1% (any fraction or multiple of 1 can be used), the number of dots in these counties or statistical units would be Glamorganshire, 46; Carmarthenshire, 6 and Anglesey, 2, corresponding with the percentage. Placing the individual dots to give a true picture of distribution is more difficult, while percentages do not often work out as whole numbers. But this method can sometimes be used to advantage to overcome the problem of 'high-value' areas. Explanation of the method used must, of course, be given in the key.

GENERAL

a. Many of the advantages of the dot map have already been mentioned: its versatility, its ability, if well-constructed, to show distribution and comparative densities, its ease of comprehension and its effectiveness in showing variations in distribution of a wide variety of commodities especially in those areas where the distribution itself is uneven. Its locational value depends to a large extent on the accuracy with which dots are placed, and on the density of the dots. It has a wide use, too, if constructed as an overlay on transparent paper for the elucidation of comparisons and correlations.

b. Dot maps, on the other hand, suffer from several disadvantages, apart from the difficulty experienced in drawing the dots and the fact that they are not easily copied. Locating the dots, for example, must reflect to a certain extent a personal, subjective decision; rarely will two dot maps, produced by two different 'composers' using the same statistics, be identical. Problems such as the placing of one dot in a widely scattered area, the fact that the area covered by the dot may not coincide with the actual area of production or population, the overcrowding of dots in 'high-value' areas, the incorrect impression that may be given by an unwise choice of dot-value or dot-size—all these defects must be borne in mind when assessing the value and accuracy of the map. These defects may be particularly acute on small scale maps, such as world or continental distribution of population; for example, using one dot to represent one million people, does one place the one dot in Liberia, or the two dots in Libya, to indicate the three million people in these countries out of a total of more than 250 or so dots for the whole of Africa? Despite all these defects, however, the dot map must remain as one of the most useful methods of representing distribution.

2. ISOLINE MAPS (Figs. 25(a), 25(b), 25(c))

One of the most widely used methods of statistical representation is the isoline, which is a line joining places or points of equal value.

Other terms used for this method are isopleths, isarithms and isometric lines but, as there is no general agreement, the word isoline (from the Greek *isos* meaning equal) is used here as being the simplest and easiest to understand. For specialised purposes, special terms are used, such as isotherms (temperature), isobar (pressure), isohyet (rainfall), isoneph (cloudiness), isobath (ocean depths), isohalines (salinity) and so on. Contour lines which, like the foregoing, are also isolines, retain the name by which they have long been known.

CONSTRUCTION

a. Statistics for as many stations as possible are plotted on the base-map in their correct places (Fig. 25(a)). As it is customary to make a tracing of the original draft on to plain paper, values and points can be marked on the base-map and used for construction purposes.

b. 'Critical values' must be carefully chosen to emphasise significant features and must take into account the extent and range of the statistics to be represented. Value intervals can be based on a regular increase, e.g. 20, 40, 60, 80 . . ., or on a geometrical progression (often used for population maps), e.g. 2, 4, 8, 16 . . ., or on natural breaks or gaps in a frequency distribution.

c. Smooth flowing lines are drawn in accordance with the intervals chosen (see Fig. 25(b)),

joining those points of equal value. If it is necessary to insert an isoline when sufficient statistics are not available on the map, interpolation is possible: the assessment of the distance of an interpolated point from two adjacent points may be calculated by joining the two points with a straight line and placing the interpolated point proportionally on that line (although in practice this is usually done by eye). For this purpose it is generally assumed that the increase or decrease between the two adjacent points is at a uniform rate.

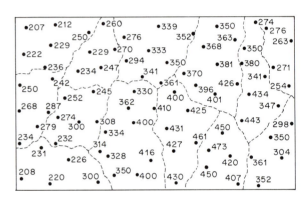

25(a) Base-map of thirteen 'counties' showing densities at various points

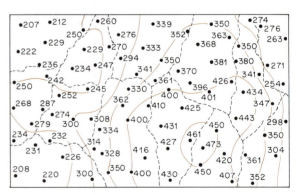

25(b) Base-map with isolines added

d. Isolines cannot, of course, cross each other.

e. Colouring or shading (see Introduction, page 1, footnote) between isolines assists in interpretation of the map and serves to draw attention to the spaces or areas between the lines rather than to the lines themselves. (See also the comments on page 2, paragraph 11).

f. Isolines should be numbered if colouring or shading is not employed (in which case a key is necessary). Values can be written either in a break in the isoline or on the line so that the top of the number is nearest to an isoline of a higher value (Fig. 25(c)). It is usually possible to use either the first or the second method without having to write figures upside down.

g. A check on the accuracy of the construction of the map may be made by making a quick cross-section; this can be done by placing a ruler or sheet of paper across the map and checking that there are no omissions in the numbering of the isolines.

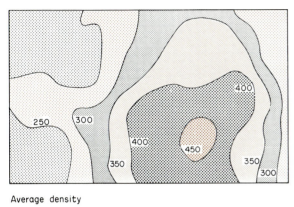

Average density

200–249	300–349	400–449
250–299	350–399	Over 450

25(c) Isoline map using part of Fig. 25(b) with 'county' boundaries removed

h. Isoline maps can also be used to represent averages, percentages or ratios. An average density figure, for example, which has been calculated for a particular area or region, may, because it is representative of the whole, be regarded as applying to the centre of that area; the figure can then be plotted in a central position within the area in the same way as individual statistics were plotted above. Obviously, the smaller the statistical units that are used in the calculation of such average densities, the more accurate will be the isoline map. As with dot maps, it is possible to make use of local knowledge in the placing of average figures. If, for example, the greatest concentration of population does not coincide with the centre of the area concerned, the figure can be placed in its correct position to indicate this.

GENERAL

In the representation of quantities and their spatial distribution, few other methods are capable of a wider range of use than isolines. Wherever it is necessary to indicate places or points of equal value, whether in absolute terms or as percentages, averages or ratios, provided that statistical information is sufficiently comprehensive, it is almost always possible to represent them by an isoline map. It is especially useful in mapping climatic data, but its use is by no means confined to this.

A comparison of Fig. 25(c) and Fig. 26(c) illustrates one of the advantages of isolines over shading (choropleth) maps. Whereas the latter method is based on the assumption that the whole statistical unit has the same average density and thus presents a generalised picture, isolines can indicate small differences within a statistical unit which may be masked by average density shading; this is particularly true if a careful choice of value intervals is made. An isoline map, too, avoids the obtrusiveness of boundary lines which are a characteristic of shading (choropleth) maps, but, even if the shading or colouring is carefully chosen, will not altogether avoid the step-like appearance of the other. Moreover, an isoline map is liable to give the incorrect impression that increase or decrease between one isoline and the next is at a uniform rate; an examination of a small-scale isoline map, such as world distribution of population, will provide evidence of this.

The fact that isoline maps can be superimposed on other types of maps without obscuring too much detail is a considerable advantage. Thus distribution of population can be superimposed on relief, yield per hectare of crops on soils, length of growing season on distribution of crops and so on. Using an overlay of tracing paper, it is then possible to determine a wide range of correlations. Sections drawn across isoline maps can be used in the same way.

3. SHADING (CHOROPLETH) MAPS (Figs. 26(a), 26(b), 26(c))

The distinction has already been made between isoline and choropleth maps. It should be noted that shading can be employed in a wide range of maps, to which the general name of 'choropleth' maps is often given. The term is derived from two Greek words—*choros*: area or space and *plethos*: multitude or number—as a reminder that such maps are quantitative areal maps, the basis of which is a relationship between quantities and area. Although the commonest form of the choropleth map is one showing density (of population, stock, etc.), it can also be used for other aspects of ratios, for averages and percentages. The term 'density', in fact, must be interpreted in its widest meaning, embracing those maps which depict average numerical values in relation to units of area.

A choropleth map, therefore, must not be used to show absolute numerical quantities, such as, for example, population figures for the counties of England and Wales derived directly from census reports.

CONSTRUCTION

a. The stages in construction are as follows:

i. If not already provided on the base map, boundaries of statistical units (e.g. county, parish, district and so on) must be drawn (see Fig. 26(b)).

ii. Calculate average densities (ratios or percentages) for each statistical unit as required (see Fig. 26(b)).

iii. Choose and draw the grades or scale of densities to be used.

iv. Indicate lightly in pencil on the map the grade of shading or colouring to be used for each statistical unit.

v. Shade or colour the map, erasing figures etc., but leaving boundary lines (see Fig. 26(c)).

b. Shading (see footnote, page 1) should show a progressive increase in density. At the lower end, a blank should be avoided as being liable to give the wrong impression (see page 2). Similarly, at the other end of the scale, solid black may draw unwarranted attention to itself (although this could be an intentional and justifiable use), prevents the inclusion of any other information and may give a false impression of an 'overcrowded' area.

c. Two different systems of shading can be distinguished—quantitative and non-quantitative. The latter, as its name suggests, does not take into account variations in density, but represents different categories by different methods of shading, colouring or indicative symbols. Examples can be found on maps depicting soils, geology, land utilisation, crop distribution and so forth. The use

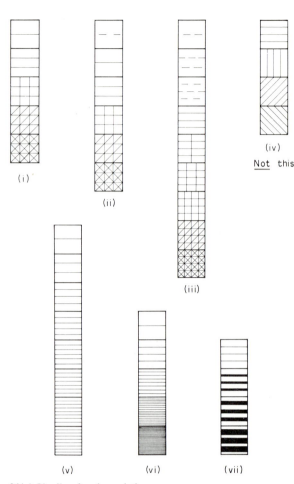

26(a) Shading for choropleth maps
  (i) 5-grade shading
  (ii) 6-grade shading
  (iii) 9-grade shading
  (iv) *Not this!*
  (v), (vi), (vii) Three types of proportional shading

of non-quantitative shading is not included here.

d. The range of values may be divided into groups by various methods. A simple arithmetical progression may be used—for example, 1–10, 11–20, 21–30, 31–40 etc.; this is useful when the range of values is not great. A geometrical progression may be employed —for example, 1–20, 21–40, 41–80, 81–160 etc.; this type of progression allows a wider range of values to be represented and at the same time gives greater consideration to the lower values. Scale values often used in density of population maps are 0–64, 64–128, 128–256, 256–512 etc. It is also possible to choose irregular intervals, but this should

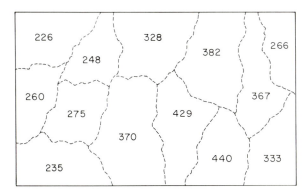

26(b) Base map of thirteen 'counties' showing average density for each. The average density figures are obtained from Fig. 25(a).

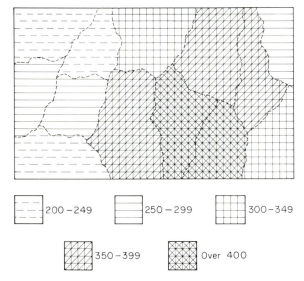

26(c) Density (shading) map based on Figs. 25(a) and 26(b).

retaining the same number of lines at a uniform distance apart but increasing the width of the line (Fig. 26(a), vii). Although a close relationship between actual density and its representation on the map can be obtained by these methods, they are difficult to draw and often difficult to interpret.

h. Do not combine shading with symbols, such as crosses, circles and the like.

i. Do not attempt to show variations in density merely by drawing lines at different angles— for example, horizontal, vertical, diagonal, as in Fig. 26(a), iv. Differences in density must be represented by differences of shading also.

j. Colouring can be used instead of shading (see Fig. 25(c)). In such cases, shades of the same colour are preferable and care must be taken to avoid 'jumping' from one colour to the next. A gradual transition from grade to grade must be aimed at.

k. A more pleasing appearance of continuity and transition is possible if shading lines are 'carried across', as far as possible, from one statistical unit to the next.

l. Use of the key can be made easier if boxes are drawn individually, as in Fig. 26(c).

m. The key should be complete, that is, all grades of shading should be included in the key, even though it may not be necessary to make use of all grades on a particular map.

n. Choropleth maps may also be used to show increases or decreases in, for example, population densities, average yields per hectare over a period of time, ratio of cultivated to uncultivated land, or any other statistical information in the form of percentages, ratios or averages. Using the same shading scheme as before for both increases and decreases, a distinction between the two can be made by shading, for instance, increases in red, decreases in blue.

only be used if the intention is to draw attention to significant irregularities in distribution; a dispersion graph (see page 16) is useful to analyse such groupings.

e. A scale composed of too few grades will give an impression of comparative uniformity, while too many grades may give a confused appearance and necessitate constant use of the key. Eight or nine would seem to form the largest practicable number of grades.

f. Examples of different methods of shading are given in Fig. 26(a), varying from 5-grade to 8- or 9-grade scales.

g. Variations in density can also be shown by proportional shading, that is, by drawing horizontal lines closer together to represent increases in density (Fig. 26(a), v–vi) or by

GENERAL

a. Great care should be taken in dividing the range of values into suitable groups or grades, and in choosing the various types of shading. Using the same statistics, it is possible to give very different impressions by altering the grades or the shading.

b. To be effective, great care must also be taken in the actual shading of the map.

c. One disadvantage of the choropleth map is that boundaries of statistical units may assume undue prominence and significance, suggesting that population densities, for example, change abruptly at the boundary.

d. Despite its appearance, it must not be assumed that densities are uniform throughout any statistical unit. This is a disadvantage inherent in the method which can be overcome to some extent by obtaining statistics for unit areas which are as small as possible; difficulty then arises in shading very small areas. Generally speaking, however, the smaller the statistical unit, the more accurate will be the map.

4. FLOW MAPS (Fig. 27)

So far we have been considering statistics concerned with quantity and time in a fixed location; we now turn to a form of statistical representation that is primarily concerned with changes in location—the flow map, the purpose of which is to represent movement from one place to another. It may show, for example, the movement of goods from area of production to market or point of export, the type and frequency of traffic by road, rail, air or water, the migratory movement of people or stock, the paths of cyclones or ocean currents, the direction, quantity or character of exports and imports and many other such statistics involving movement. The direction of flow is indicated by a line, the amount by the width of the line and the character usually by colouring or shading of various types. It is important to remember that this is not an areal representation; the width of the flow-line is the important factor—the length of the flow-line shows the direction of movement and the distance involved but has no significance as far as quantities are concerned.

CONSTRUCTION

a. The first consideration should be the width of the flow-line, which must be proportional to the quantity or value of the commodity to be represented. Extremes of width—too great or too small—should be avoided.

b. The flow-line can be drawn in several ways: it can be a solid line of varying width, drawn to scale (reference to a key is necessary for its exact interpretation); it can be shown by a line chosen from a series of graduated band-widths, each band-width representing a range of values within the total range (see Fig. 27); it can also be drawn as a series of parallel lines of uniform width, each line representing a fixed quantity.

c. In drawing the flow-line on the base-map, it is not necessary to follow all the twistings and windings of a road or railway. The line is superimposed on the route, sharp angles and bends being 'rounded off' to give a generalised impression of the route followed.

d. Junction of flow-lines presents little difficulty, except when the flow-line has been sub-divided to show various components (for example, of traffic). In such cases it may be necessary to break the main flow-line.

e. The meeting place of various flow-lines (for example, a market town, bus station, port of import or export, collecting centre and so on) can be shown by various methods. The easiest method is to bring all lines to meet—this will result in an irregular shaped nucleus, but, as we are concerned with the movement of commodities and not representation of meeting places, this is acceptable. The meeting place of flow-lines may also be represented by a circle of any convenient size, although often an attempt is made to indicate the size or importance of the centre by varying sizes of circles.

f. The terminal collecting point of several flow-lines may also be represented as a square or circle, either of which may be sub-divided (as a divided rectangle or pie chart) to show the total values of individual components moving towards the collecting centre. This information, however, is better represented by individual located pie charts which do not form part of the flow-lines.

g. Colour can be used to great advantage to represent different categories of commodities, to distinguish imports and exports or generally to supply further information.

h. When an attempt is made to sub-divide the flow-line into its various components, difficulty often arises in representing the smaller amounts. In such cases it is advisable to draw an inset at an exaggerated scale (see Introduction, page 2).

i. Factual information may also be written on the flow-line or alongside it.

GENERAL

This method of representing flow or movement has a wide variety of uses of practical value. It is, indeed, the only method of representing movement in common use and is capable of many refinements and modifications.

Its disadvantage lies in its lack of immediate or exact interpretation, although it gives a visual impression which is realistic and effective, while statistical information may be added on the map.

Difficulty, too, arises in drawing parallel 'double-track' flow-lines that remain parallel to each other along curved sections of the route, but some assistance in overcoming this difficulty may be found in marking the flow-line with a pair of dividers or a compass and pencil, or, more simply, by converting curves into a series of straight lines, which preserve the general direction of movement.

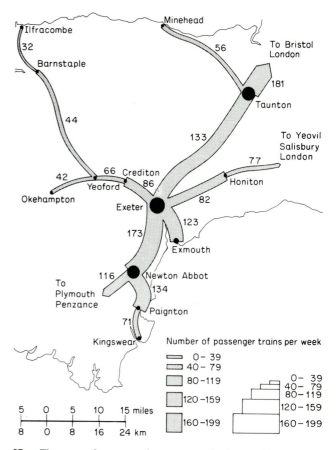

27   Flow map: frequency of passenger trains (per week) from Exeter, April 1970

Exercises

NOTE

1. Before attempting any of the following exercises you are advised to read the Introduction again.

2. The exercises are planned to give practice in almost all the types of statistical maps and diagrams described in this book. When working through them, do not resort to using one type of representation only, e.g. bar graph or pie chart, but gain practice and experience in as wide a variety of methods as you can, provided always that the method chosen is the most suitable for the purpose.

3. Calculations should always be shown—not as a jumbled and meaningless mass of figures, but set down neatly with a clear explanation of each step in your calculation.

4. In some exercises, instructions are given concerning the method of representation to be used; in others, you must choose the best method, remembering that, of several methods that may be available, one particular method is usually the most effective in representing the statistics provided.

EXERCISE 1.

Construct a statistical graph or chart to illustrate the following figures which refer to Britain's estimated public expenditure, 1970–1971.

	£ million 1970–1971 (estimated)
Social Security	3,528
Education	2,474
Defence budget	2,281
Health and Welfare	1,958
Housing	1,126
Technological services and other assistance to employment and industry	1,088
Local environmental services	887
Roads and public lighting	698
Law and order	658
Agriculture, fisheries and forestry	400
Overseas aid and other overseas services	363
TOTAL*	£16,926

(Source: U.K. Central Office of Information.)
* See Introduction, p. 2, Section 14.

EXERCISE 2.

Draw a series of compound (or divided) bar graphs to represent the following statistics which relate to the exports from the port of Cotonou ('000 tons) from 1960 to 1967.

	Palm Kernels	Palm Oil	Ground-nuts	Cotton	Total exports
1960	61·3	15·4	12·9	2·2	104·6
1961	41·5	11·5	12·5	1·3	93·4
1962	43·9	9·3	4·3	0·6	75·0
1963	50·5	9·2	6·5	1·4	83·2
1964	56·1	12·7	3·9	1·1	89·7
1965	14·3	13·2	15·6	1·2	93·9
1966	8·4	12·3	29·5	2·7	103·9
1967	4·8	7·6	82·49	3·5	157·2

(Source: Direction du Port de Cotonou).

EXERCISE 3.

The following statistics summarise the export trade of Abidjan, 1960 to 1966. Figures refer to percentages by value of total exports.

Draw a compound line graph to represent these statistics.

	Coffee	Cocoa	Timber	Bananas/ Pine- apples	Palm Kernels	Minerals	Others
1960	50	23	16·0	5·2	1·6	1·4	2·8
1961	46	22	19·0	6·2	0·8	3·0	3·0
1962	38	22	19·5	7·5	0·5	2·3	10·2
1963	38	19	21·0	7·8	0·4	1·4	12·4
1964	43	20	24·0	5·8	0·4	2·0	4·8
1965	38	16	27·0	6·0	0·4	1·6	11·0
1966	39	16	24·0	5·9	0·3	2·2	12·6

(Source: Direction du Port d'Abidjan).

EXERCISE 4.

The following statistics refer to the production of cereals in France in 1969 (final figures) and 1970 (provisional figures) in '000's metric quintals.

Column 3 indicates the variation in production. Construct bar graphs to illustrate these figures.

	1969	1970	Variation
Soft Wheat	144,460·6	127,878·4	− 11·7%
Durum	3,598·1	4,768·2	+ 33·4%
Barley	96,258·6	81,629·4	− 15·3%
Rye	3,314·5	3,047·5	− 2·4%
Oats	23,416·9	21,047·2	− 10·3%
Maize	57,629·6	74,720·5	+ 29·7%
Rice	975·9	1,026·9	+ 5·4%
Sorghum	1,984·1	1,661·0	− 16·3%

1 metric quintal = approx. 100 kg

(Source: Statistics Service, Ministry of Agriculture, Paris).

EXERCISE 5.

Prepare 3 outline maps of Australia, marking the states. Using the information below:

a. *Calculate the density of population per 100 square miles in the states of (i) Victoria and (ii) Northern Territory.*

b. *Draw located proportional circles representing the total population of each state.*

c. *Draw located bar graphs representing the male and female population of each state.*

d. *Draw a series of proportional squares to represent the areas of individual states.*

e. *Construct a divided rectangle to represent both the total population and the population of the individual states.*

f. *Construct a large pie chart to represent total area and area of states.*

g. *Complete a density (shading) map of the continent by states.*

State	Area (sq mls)	Males (pop.)	Females (pop.)	Total Pop. (1969)	Density (per 100 sq mls)
New South Wales	309,433	2,247,000	2,227,800	4,474,800	1,446
Victoria	87,884	1,698,700	1,685,400	3,384,100	—
Queensland	667,000	895,900	872,100	1,768,000	265
South Australia	380,070	575,000	569,000	1,144,000	301
West Australia	975,920	481,800	464,500	946,300	97
Tasmania	26,383	196,000	192,400	388,400	1,473
N. Territory	520,280	36,800	31,200	68,000	—
Australian Capital Territory	939	62,800	59,700	122,500	12,993
Total	2,967,909	6,194,000	6,102,100	12,296,300	414

(Source: *Statesman's Year Book*, 1970–1971).

EXERCISE 6.

Draw a map of England, Scotland and Wales. On it mark the towns listed below and choose one method to represent the relative populations. What other methods could you use and what are the advantages and disadvantages of each method?

Town	Population (thousands) 1961
Bristol	437
Cardiff	257
Glasgow	1,055
Coventry	306
Edinburgh	468
Greater London	7,997
Kingston upon Hull	303
Leicester	273
Liverpool	746
Nottingham	312
Sheffield	494
Manchester	662
Stoke on Trent	265
Newcastle upon Tyne	270
Birmingham	1,107
Wolverhampton	262
Bradford	296
Leeds	511

(Source: U.K. Central Office of Information).

EXERCISE 7.

Draw a large pie chart to illustrate the following statistics (exports from Britain by value, 1969).

Exports	Value (£ million)
Machinery (other than electrical)	1,418
Chemicals	685
Passenger cars, commercial vehicles and chassis	497
Electrical machinery	466
Textile manufactures	347
Non-ferrous metals	312
Iron and steel	285
Miscellaneous metal manufactures	215
Food and live animals	191
Scientific instruments	179
Aircraft and parts	175
Whisky	167
Petroleum products	143
Total	7,338

(Source: U.K. Central Office of Information).

EXERCISE 8.

Construct a statistical chart or graph to illustrate the following table which shows the percentage of total British food supplies provided by home agriculture in 1967 as compared with the pre-war average.

Food Product	Pre-War Average %	1967 %
Wheat and flour	23	50
Oils and fats	16	12
Sugar	18	30
Carcase meat & offal	51	72
Bacon & ham	32	34
Butter	9	8
Cheese	24	43
Shell eggs	71	98
Milk for human consumption	100	100
Potatoes	96	95

(Source: U.K. Central Office of Information).

EXERCISE 10.

Of the total Swiss population (5,429,061) in 1960, 4,844,322 were Swiss citizens and 584,739 were foreigners. In that year 2,512,411 were classified as employed or self-employed persons; of this total 1,239,009 were engaged in industry and crafts, 346,215 in commercial banking and insurance, 280,191 in agriculture and forestry and 248,634 in hotel and restaurant business and transport. Draw statistical diagrams to illustrate these figures.

EXERCISE 9.

Show by means of a statistical chart or diagram the following information which indicates percentage of Swiss population by mother tongue (1960). Column A refers to percentage of Swiss citizens, Column B to total Swiss population.

Mother Tongue	A (%)	B (%)
German	74·4	69·3
French	20·2	18·9
Italian	4·1	9·5
Romansch	1·0	0·9
Others	0·3	1·4

EXERCISE 11.

Illustrate by means of a statistical chart or diagram the following figures which show the utilisation, by percentage, of cultivated agricultural land in Norway, 1968 :

Meadows for hay, etc.	48%
Grain and peas	23%
Meadows for pasture	15%
Potatoes	4%
Fruits and vegetables	3%
Other open-land crops	4%
Unused and fallow	3%

EXERCISE 12.

Draw pie-charts (divided circles) to illustrate the following statistics :

a. *Distribution of working population in Netherlands by percentage.*

Year	Agriculture and fisheries	Industry (including mining)	Trade and transport	Other professions
1890	33	32	16	19
1920	24	39	21	16
1960	11	42	23	24
1967	8	42	24	26

b. *Distribution of working population in other countries by percentage (1967).*

Country	Agriculture and fisheries	Industry (including mining)	Trade and transport	Other professions
Gt. Britain	2	48	23	27
U.S.A.	5	34	30	31
Japan	24	32	28	16
Mexico	49	21	14	16

EXERCISE 13.

The table below shows the areas and populations of the eight standard administrative regions of England in 1961 together with those of Wales, Scotland and Northern Ireland for the same year. Column 3 indicates the estimated (projected) population of these regions in 1981.

a. *Using the statistics in Columns 1 and 2 and a tracing of the map (Fig. 28) on page 56, prepare*

a map to illustrate the relative densities of population (by regions).

b. *Using information in Column 3, construct on another tracing a map to show projected population changes over the period 1961–1981. Care should be taken to devise a method of presentation which will give an accurate impression of the comparisons between the regions.*

Standard Regions of England:	Area in sq mls	Thousands	
		1961	1981 (projected)
North	7,470·9	3,249·0	3,570
Yorkshire & Humberside	5,481·8	4,628·1	5,213
North-West	3,083·3	6,546·6	7,401
East Midlands	4,701·1	3,107·0	3,860
West Midlands	5,024·4	4,762·9	5,699
East Anglia	4,851·6	1,489·8	2,008
South-East	10,584·8	16,350·5	18,585
South-West	9,134·9	3,436·1	4,134
Wales	8,015·8	2,635·2	2,923
Scotland	29,796·1	5,183·8	5,314
Northern Ireland	5,451·8	1,427·4	1,664

(Source: U.K. Central Office of Information).

EXERCISE 14.

The following information (1968) refers to New Towns in East and South-East England. Trace two outlines of the map (Fig. 29) on page 56; use one to

represent the statistics in Column 1 below and the other to represent the statistics in Columns 2 and 3.

Name of New Town	Area designated (hectares)	Population End 1967	Ultimate Population
Stevenage	2,500	61,700	105,000
Crawley	2,420	63,700	75,000
Hemel Hempstead	2,400	67,900	80,000
Harlow	2,560	75,800	80,000
Hatfield	940	24,700	29,000
Welwyn	1,700	44,300	50,000
Basildon	3,120	75,000	140,000
Bracknell	1,320	28,300	60,000
Corby	1,720	47,500	80,000
Milton Keynes	8,800	40,000	250,000
Peterborough	6,400	81,000	175,000
Northampton	8,000	131,000	220,000

(Source: U.K. Central Office of Information).

28   Administrative regions of the British Isles

29   Southeast England new towns

EXERCISE 15.

Prepare a large map of Brazil and on it mark the towns listed below. Using a method of your own choice, show on the map the relative populations of the towns.

Town	Population (1970 census—provisional)
São Paulo	5,901,533
Rio de Janeiro	4,296,782
Belo Horizonte	1,232,708
Recife	1,078,819
Salvador	1,000,647
Porto Alegre	885,567
Fortaleza	842,231
Belem	642,514
Curitiba	603,227
Brasilia	544,862

(Source: *Europa Year Book*, 1971).

EXERCISE 16.

The best way to understand a semi-logarithmic graph is to draw and construct your own (see page 18). Using the statistics below, draw semi-logarithmic graphs to show (a) the value in £L '000 of exports of petroleum from Libya 1961 to 1968; (b) the tonnage (long tons) of iron ore shipped from Monrovia 1950 to 1967.

a. Exports of Petroleum from Libya, 1961–1968
(£L '000)

1961	4,097
1962	46,984
1963	116,861
1964	216,400
1965	280,331
1966	351,441
1967	416,426
1968	667,262

(Source: Census & Statistical Dept., Tripoli).

b. Export of Iron Ore from Monrovia (long tons)
1950–1967

1950	
1951	170,000
1952	888,140
1954	1,190,050
1956	2,037,450
1958	2,045,370
1960	3,037,540
1962	3,643,000
1964	5,649,480
1965	7,152,760
1966	8,433,760
1967	9,394,450

What is the purpose of a semi-logarithmic graph? How does it differ from a simple line graph? Illustrate your answer by re-drawing the statistics in (a) as a simple line graph.

EXERCISE 17.

Using a tracing of the map of Aquitaine (Fig. 30) on page 57 and the information below, construct a population density (shading) map of the Departments of Aquitaine. What other information would you require if you were asked to construct a distribution of population (dot) map?

Department	Area (sq km)	Population 1962 Census
Gironde	10,726	935,448
Landes	9,364	260,495
Basses-Pyrénées	7,712	466,038
Dordogne	9,224	375,455
Lot et Garonne	5,385	275,028

EXERCISE 18.

Draw a divided rectangle to illustrate the following statistics relating to land use in the Netherlands (1969):

Grassland	46%
Arable land	26%
Woods, heaths and dunes	13%
Buildings and roads	10%
Horticulture	5%

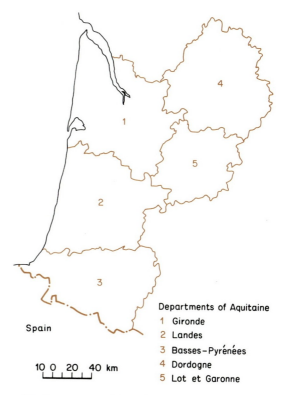

Spain

10 0 20 40 km

Departments of Aquitaine
1 Gironde
2 Landes
3 Basses-Pyrénées
4 Dordogne
5 Lot et Garonne

30 The Departments of Aquitaine

EXERCISE 19.

Draw a flow-line diagram to illustrate the export trade (1968) of Uruguay, assuming that all exports pass through Montevideo. Figures are in $'000.

United Kingdom	37,908
U.S.A.	21,699
Italy	12,982
Spain	12,115
West Germany	11,674
Netherlands	10,533
Brazil	7,455

(Source: *Europa Year Book, 1971*).

EXERCISE 20.

Using a tracing of the map of Uganda and the flight time-table (Fig. 31 over), construct a flow-line diagram to illustrate the frequency and direction of Uganda's internal air services.

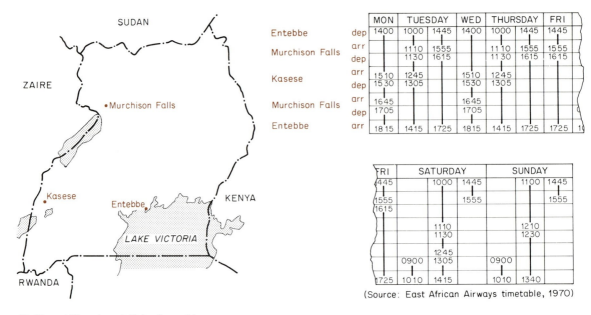

Map labels: SUDAN · ZAIRE · • Murchison Falls · • Kasese · Entebbe · KENYA · LAKE VICTORIA · RWANDA

Flight time-table:

		MON	TUESDAY		WED	THURSDAY		FRI
Entebbe	dep	1400	1000	1445	1400	1000	1445	1445
Murchison Falls	arr		1110	1555		1110	1555	1555
	dep		1130	1615		1130	1615	1615
Kasese	arr	1510	1245		1510	1245		
	dep	1530	1305		1530	1305		
Murchison Falls	arr	1645			1645			
	dep	1705			1705			
Entebbe	arr	1815	1415	1725	1815	1415	1725	1725

		FRI	SATURDAY			SUNDAY		
Entebbe	dep	1445		1000	1445		1100	1445
Murchison Falls	arr	1555			1555			1555
	dep	1615						
Kasese	arr			1110			1210	
	dep			1130			1230	
Murchison Falls	arr			1245				
	dep		0900	1305		0900		
Entebbe	arr	1725	1010	1415		1010	1340	

(Source: East African Airways timetable, 1970)

31 Map of Uganda and flight time-table

Appendix

METRIC CONVERSION TABLES

LENGTH

Centimetres		Inches	Metres		Yards	Kilometres		Miles
2·540	1	0·394	0·914	1	1·094	1·609	1	0·621
5·080	2	0·787	1·829	2	2·187	3·219	2	1·243
7·620	3	1·181	2·743	3	3·281	4·828	3	1·864
10·160	4	1·575	3·658	4	4·374	6·437	4	2·485
12·700	5	1·969	4·572	5	5·468	8·047	5	3·107
15·240	6	2·362	5·486	6	6·562	9·656	6	3·728
17·780	7	2·756	6·401	7	7·655	11·265	7	4·350
20·320	8	3·150	7·315	8	8·749	12·875	8	4·971
22·860	9	3·543	8·230	9	9·843	14·484	9	5·592
25·400	10	3·937	9·144	10	10·936	16·093	10	6·214

1 cm = 0·393701 in
1 in = 2·54 cm

1 m = 1·09361 yd
1 yd = 0·9144 m

1 km = 0·621371 mile
1 mile = 1·60934 km

AREA

Square Centimetres		Square Inches	Square Metres		Square Yards	Hectares		Acres
6·452	1	0·155	0·836	1	1·196	0·405	1	2·471
12·903	2	0·310	1·672	2	2·392	0·809	2	4·942
19·355	3	0·465	2·508	3	3·588	1·214	3	7·413
25·806	4	0·620	3·345	4	4·784	1·619	4	9·884
32·258	5	0·775	4·181	5	5·980	2·023	5	12·355
38·710	6	0·930	5·017	6	7·176	2·428	6	14·826
45·161	7	1·085	5·853	7	8·372	2·833	7	17·297
51·613	8	1·240	6·689	8	9·568	3·237	8	19·769
58·064	9	1·395	7·525	9	10·764	3·642	9	22·240
64·516	10	1·550	8·361	10	11·960	4·047	10	24·711

$1 \text{ cm}^2 = 0·155 \text{ in}^2$
$1 \text{ in}^2 = 6·4516 \text{ cm}^2$

$1 \text{ m}^2 = 1·19599 \text{ yd}^2$
$1 \text{ yd}^2 = 0·836127 \text{ m}^2$

1 ha = 2·47105 acres
1 acre = 0·404686 ha

WEIGHT

Kilograms		Av. pounds	Metric Tonnes		Long Tons
0·454	1	2·205	1·016	1	0·984
0·907	2	4·409	2·032	2	1·968
1·361	3	6·614	3·048	3	2·953
1·814	4	8·819	4·064	4	3·937
2·268	5	11·023	5·080	5	4·921
2·722	6	13·228	6·096	6	5·905
3·175	7	15·432	7·112	7	6·889
3·629	8	17·637	8·128	8	7·874
4·082	9	19·842	9·144	9	8·858
4·536	10	22·046	10·161	10	9·842

1 kg = 2·20462 lb
1 lb = 0·45359237 kg

1 t = 0·984207 ton
1 ton = 1·01605 t

VOLUME

Litres		Gallons
4·546	1	0·220
9·092	2	0·440
13·638	3	0·660
18·184	4	0·880
22·730	5	1·100
27·276	6	1·320
31·822	7	1·540
36·368	8	1·760
40·914	9	1·980
45·460	10	2·200

1 l = 0·219969 gal
1 gal = 4·54609 l

VELOCITY

m.p.h.		knots
1·15	1	0·87
2·30	2	1·74
3·46	3	2·60
4·61	4	3·48
5·76	5	4·35
6·91	6	5·21
8·06	7	6·08
9·22	8	6·95
10·37	9	7·82
11·52	10	8·69

1 m.p.h. = 0·8684 knots
1 knot = 1·152 m.p.h.

m.p.h.		m/sec (m.s^{-1})
0·447	1	2·237
0·894	2	4·474
1·341	3	6·711
1·788	4	8·948
2·235	5	11·185
2·682	6	13·422
3·129	7	15·659
3·576	8	17·896
4·023	9	20·133
4·470	10	22·370

1 m.p.h. = 2·237 m/sec (m.s^{-1})
1 m/sec (m.s^{-1}) = 0·447 m.p.h.